During 1906 the sub-basement of the Boston Store on Madison was excavated down to tunnel grade so that coal, cinder and general freight traffic could be handled by train. A construction train is seen in the still-unfinished sub-basement.

Larry Best Collection

THE CHICAGO TUNNEL STORY

EXPLORING THE RAILROAD "FORTY FEET BELOW"

By Bruce G. Moffat

Bulletin 135 of the Central Electric Railfans' Association

THE CHICAGO TUNNEL STORY

EXPLORING THE RAILROAD "FORTY FEET BELOW"

By
Bruce G. Moffat

Bulletin 135 of the Central Electric Railfans' Association

An Illinois Not-for-Profit Corporation
Post Office Box 503, Chicago, Illinois 60690, U.S.A.

The Chicago Tunnel Story was designed by Jack Sowchin.

CERA Bulletins are technical, educational references prepared as historic projects by members of the Central Electric Railfans' Association, working without salary due to their interest in the subject. This bulletin is consistent with this stated purpose of the corporation: To foster the study of the history, equipment and operation of electric railways.

ISBN: 0-915348-35-7

Preceding page:
The wide angle lens used by the photographer makes this newly constructed tunnel look gigantic. In reality the bore was only 7 feet, 6 inches high while the track is part of a temporary 14 inch gauge construction railroad. The year is 1904.

Larry Best Collection

Table of Contents

Foreword

For more than half a century, Chicago possessed the best-kept secret in the annals of railroad transportation. Alone among the great cities of the world, the Windy City was served by a subterranean freight railway system that spanned the length and breadth of its central area – the famed "Loop" district. The existence of this unique system was not a total secret, of course. It was simply the result of its being hidden from easy public view – forty feet below street level.

Although confined to a geographic area of less than four square miles, this was a system that at its peak utilized nearly 150 electric locomotives operating over nearly 60 miles of two-foot-gauge track. At the time of its construction just after the turn of the century it was considered a major engineering accomplishment with seemingly unlimited profit-making potential. At the same time, its construction and operation garnered surprisingly little public attention. Unfortunately, it failed to attract sufficient customers to ensure long term profitability.

Over the years, the author has read with interest the many books and articles written about two-foot-gauge railroads in the United States. Most of these works focused on the several steam powered railroad companies that operated in the state of Maine. The largest of these was the Sandy River & Rangley Lakes Railroad which, at its peak, had more than 100 miles of route and remained in operation until 1935; this was just long enough for it to be discovered and photographed by rail enthusiasts. By comparison, precious few historical accounts were written about the next largest of the two-footers in terms of mileage: the Illinois-based Chicago Tunnel Company, which did not cease operations until 1959.

Aside from sharing a common track gauge and being located in the "snow belt," the Sandy River and the Chicago Tunnel Company had almost nothing in common. Virtually all of the Tunnel Company's trackage was in tunnels mined beneath the streets of a major city; this fact alone made it unique among the world's freight railways. It also meant that maintaining regular operations during periods of severe winter weather was not a problem. The Sandy River was not as blessed, sometimes having trains stranded for days at a time due to blizzards.

The existence of this long-idle, but largely intact, railroad lying just forty feet below the downtown thoroughfares of one of America's largest cities caught the interest of this author while attending college in the mid-1970's. Fueled by occasional "filler" articles in the local press, I began to research the story of this most interesting, and unusual, of railroads. While the research and preparation of any book can be considered a challenge, the research phase included one goal that, for a while seemed out of reach: gaining access to the tunnels themselves. Available points of entry were few and obtaining permission to enter the tunnels for research purposes was, to say the least, difficult. Most of the tunnels were by that time controlled by the City of Chicago. The city, understandably, was reluctant to divert personnel from more important tasks to escort the author on walking trips through the bores.

A few downtown buildings still had access to the tunnels, or even owned short sections of the bores, unfortunately, information of these locations was difficult to come by. But as luck would have it, my father, James G. Moffat, managed the night security staff at Marshall Field's downtown store, which coincidentally had never bothered to sever its connection with the tunnel network. These circumstances enabled me to make my first visit the railroad "fort feet below" in 1974.

Eventually, arrangements were made with the city that allowed for extensive inspection of the tunnels that contributed greatly to my understanding of this very unusual railroad. The results of my research were first published under the title *Forty Feet Below - the Story of Chicago's Freight Tunnels* in 1982. In succeeding years additional information about the company's early years was discovered. This, along with the desire to update the book to include some recent events, including the extraordinary "Loop Flood" of 1992, resulted in the preparation of this expanded volume about the railroad "forty feet below."

Bruce G. Moffat

Ron T. Kuziel

On June 18, 1974, author Bruce G. Moffat (at left) had just completed his first year of college when he made his first visit to the half-forgotten freight tunnel system. Standing next to him in Marshall Field's siding under Wabash Avenue are (left to right): Andris J. Kristopans, Phillip F. Cioffi, and James G. Moffat. Not pictured is Ron T. Kuziel who recorded this auspicious moment on film.

Acknowledgments

The preparation of this revised and enlarged treatment of the Chicago freight tunnel system would have been impossible had it not been for the help of many interested individuals and organizations who assisted the author in locating materials, offered encouragement, or assisted in the research phase.

During the preparation of the original volume in the early 1980s, I was assisted by a number of individuals including Joseph Corona and Julian Waisnor who were among the last employees of the Chicago Tunnel Company and provided many details on daily operations. Also of considerable assistance was James Lyons, manager of the Pittsfield Building for several years. Mr. Lyons allowed the author to inspect the Pittsfield's connection to the tunnel and also provided details about the never-used connection to the Prudential Building that would have otherwise gone unrecorded. Norman Radtke, physical plant manager and Leonard Carrion, chief engineer, of the Field Museum of Natural History were instrumental in providing data on that institution's involvement with the freight tunnels.

The City of Chicago's Municipal Reference Library, the Chicago Historical Society, the Commonwealth Edison Company Library, the Chicago Transit Authority's Anthon Memorial Library, the Chicago Public Library, and the John Crerar Library at the University of Chicago were instrumental in locating company promotional brochures, nearly century-old trade journals and other research materials. The Archives Department of the Chicago Tribune Company made available its extensive news clippings file, which shed additional light on the last decade of operations as well as the experimental mail trains in the 1950s. Telephone historian and collector Larry Best allowed the author access to an album of construction photographs and a scrapbook of early news clippings about the system's construction. Without Larry's contribution, the expanded treatment of the tunnel system's early years would have been almost impossible. Telephone historian Stan Swihart provided additional information on the early telephone operations. J.J. Sedelmaier loaned the author some early promotional booklets concerning the system's construction.

Information on the system's rolling stock was assembled from a number of sources. The fleet lists were developed based on company documents on file with the Illinois and Interstate Commerce Commissions, surviving builders' records and other sources, and were reviewed by Phillip O'Keefe and Mark Landgraf. Mark also supplied valuable information on the Baldwin-Westinghouse units and the one-of-a-kind Whitcomb. Information on the third rail Morgan cog locomotives was provided through the courtesy of the Goodman Equipment Corporation (successor to the Goodman Manufacturing Company). J.W. Brantner, vice president of development of the Jeffrey Mining Machinery Division of Dresser Industries, supplied valuable data on the locomotives supplied by his company. He also was able to locate a number of never before published photographs, as well as locomotive

Bruce G. Moffat

Left behind during salvage operations two decades earlier, ash car 822 was found by the author at the intersection of the Orleans and Hubbard tunnels in 1980.

plans and specification sheets.

Since few rail enthusiasts were allowed to visit the tunnels, the number of photographs available was quite limited. Surviving original prints of company-sponsored photos were also in short supply, necessitating the reworking of some pictures originally published in promotional booklets and early newspapers. Ted Koston and David McNamara performed minor dark room miracles in this regard. Jack-Ad Graphics provided computer scanning and book design services. Research assistance was also provided by Samuel Polonetzky, Roy G. Benedict, Melvin Bernero, Andris Kristopans, William Shapotkin, Vince Dawson, Edward Anderson Jr. and Zennon Hansen.

Thanks are also due to Marisue O'Connor who reviewed the manuscript and made various editorial corrections and suggestions.

Lastly, the author extends his thanks to the staffs of Chicago Department of Public Works and the Chicago Department of Transportation for allowing access to the tunnels for research purposes over the years and providing information on their current role as a utility conduit.

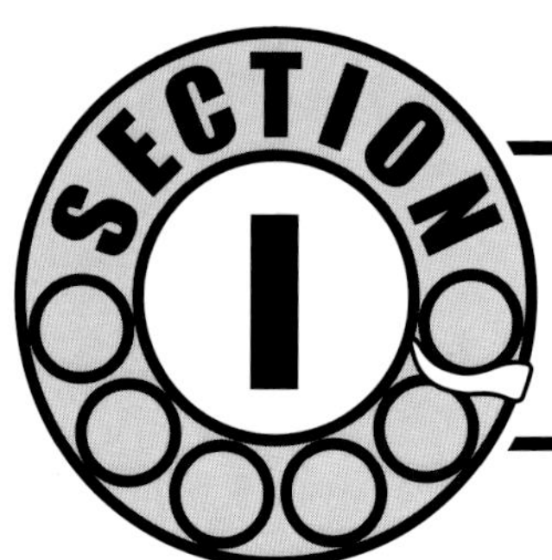

The Illinois Telephone & Telegraph Company

1 Telephones and Politics

Chicago's one-of-a-kind subterranean railway started out at the turn of the century as something quite different: an underground telephone system – at least that is what its promoters told the city fathers while they mined beneath the congested streets of the city's "Loop" business district.

One of the earliest references to this project to be found in the public records appeared on June 27, 1898. On that date Chicago Alderman Edward J. Novak of the 8th Ward introduced a proposed ordinance in the City Council to grant a franchise to an unidentified company to construct and operate a telephone system. This enterprise would be in competition with the already entrenched Chicago Telephone Company, a predecessor of the Illinois Bell Telephone Company (today's Ameritech).

The proposal called for a 50-year franchise and stipulated that all cables installed in the area bounded by North Avenue, Ashland Avenue, 39th Street and Lake Michigan be placed underground in conduits. Cables installed outside of this approximately 16 square-mile area could be placed above ground for a 15-year period, after which time they too would have to be buried. Wires could be brought above ground within each block to reach subscribers in an area not to exceed four blocks from the underground connection. One clause even required that within six months of placing its telephone plant in operation, the holder of this franchise would have to install at least one free telephone in each of several city offices and departments.

The proposed ordinance was referred to the Committee on Gas, Oil and Electric Light for consideration. In a May 20, 1899, article recapping local telephone developments, the Chicago *Tribune* reported that the heretofore unnamed backers of this initiative were a group of wealthy St. Louisans, headed by brewer Adolphus Busch. Their interests were represented by former U.S. Solicitor General Charles H. Aldrich (who handled the franchise negotiations with the aldermen), H. J. Hanford, and Milo G. Kellogg. The article went on to say:

> The St. Louis people who, with Mr. Hanford as the guide, have built a telephone exchange in that city competing with the Bell company, and exchanges in other cities, united in a telegram to the [Chicago] authorities relating their intentions to build immediately. But operations didn't begin.
>
> During the recent Mayoralty campaign ex-Governor Altgeld said on the stump the ordinance had been sold and intimated the purpose of the promoters in getting it was simply to find a buyer. It is said in financial circles that the ex-Governor's assertion was partly correct in that while the ordinance hadn't been sold it was in the market. The explanation of this is stated to be that Mr. Busch and his friends found a different state of affairs here from what was expected. The stockholders of the Chicago Telephone company include many of the large capitalists, bankers, and brokers. Many of them did not desire to take part in any enterprise which tended to depreciate the market value of Chicago Telephone stock and they declined to become stockholders in the rival company... At any rate, there seems little doubt now that the ordinance has been for sale almost from the time of its passage, and yesterday the rumors culminated around the name of A. G. Wheeler.

Left: *Early Stroger-type dial phone.*
GTE Automatic Electric

Capitalist Albert G.Wheeler was quoted as saying that he had the ordinance and that he had already raised the money necessary to install the system, but that operations would not begin until 10,000 subscribers had been signed up. No mention was made of what it had cost to acquire the franchise, although one railroad industry publication put the tab at a hefty $1 million. His fellow investors were mostly Chicagoans and apparently did not include anyone from the St. Louis group.

Incorporation

On June 23, 1898, four days prior to Alderman Novak's introduction of the proposed ordinance, Wheeler's group filed an application for incorporation with the Illinois Secretary of State's office in Springfield. In its application for incorporation, the Illinois Telephone & Telegraph Company (IT&T) stated that its corporate objectives were to "manufacture, buy, sell and deal in telephone instruments, telegraph instruments and all kinds of electrical appliances and apparatus; to buy, sell, lease and deal in electrical

Albert G. Wheeler

Prominent among the IT&T's backers was Albert G. Wheeler, who was introduced to the Chicago press variously as the builder of an underground conduit trolley system in Washington, D.C. and as a "spectacular eastern traction and utility magnate."

Born in New York City on April 27, 1854, Wheeler was educated in the local public school system before enrolling at New York College in 1868. Entering the commission business at the age of 18, he eventually accumulated sufficient capital to purchase an interest in the produce transportation business on the Hudson River. His business flourished and he was soon prominent in the lighterage (maritime) transportation business in New York harbor. He apparently also had some involvement in the promotion of the conduit current collection system for street railways in New York and Washington, D.C.

Wheeler's move to Chicago apparently resulted from a business trip as related in this well-embellished account that appeared in *History of Cook County Illinois*, published by the Goodspeed Historical Association in 1909:

> While present in [Chicago] on a business trip, the sad state of the city's freight transportation facilities appealed to him with that irresistible fascination which great difficulties always seem to exercise over genius... His plan was for a network of tunnels connecting the shipping departments of the large mercantile houses with the freight yards of the railroad companies. Securing a franchise from the city he organized the Illinois Telephone and Telegraph company, and obtained for the company the right to establish a system of "sounds, signals and intelligence, by electricity or otherwise" and to run conduits under all the streets and alleys and

Bruce G. Moffat Collection

Tunnel promoter and president Albert G. Wheeler had already left the enterprise when his portrait appeared in a 1910 publication about Chicago's business leaders.

patents and patent rights; also to acquire, construct, build, buy, sell, lease and operate telephone plants, telegraph plants and plants for the conveyance and transmission of sound and signals by electricity." The IT&T's organizers were identified as Ralph W. Bowman, Samuel H. McLaughlin and Thomas J. Holmes. The IT&T was capitalized at a relatively modest $250,000 and its incorporation became effective on July 6 of that year.

Following incorporation, the company launched a publicity campaign to alert the citizens of Chicago that they would soon be the beneficiaries of one of the first completely automatic dial-type telephone systems in the country (or for that matter, the world). This system was touted to be far superior to the conventional operator-assisted manual system operated by the Chicago Telephone Company.

The Automatic Electric Connection

One of Wheeler's close business associates was Charles H. Aldrich. Besides having represented the Busch group, Aldrich was the legal counsel for the Strowger Automatic Telephone Exchange. Almon B. Strowger and Joseph Harris had founded this Chicago-based telephone equipment manufacturing company in 1891 to promote the use of an automatic telephone switching system that Strowger had developed in 1889. The first city to use the Strowger system was LaPorte, Indiana, in 1892. This pioneer installation utilized push-button equipped telephones rather than an operator to complete telephone calls. Strowger's revolutionary system was also successfully demonstrated

> under the Chicago river and its branches. "Intelligence" was taken to cover newspapers and mail matter and it was argued by Mr. Wheeler, that a tunnel which would accommodate cars large enough to handle such matter could also carry merchandise. These cars, he decided, should be large enough to take in the largest box or package that could be put through the doors of a railroad freight car. They should also be of such a size and weight that they could enter any mercantile building and be raised by elevators to any floor...

If this account is indeed accurate, it is interesting to note that the franchise application carefully avoided any references to railroads or the transportation of goods. In fact, all representations about the nature of the company's planned business activities were confined to the installation and operation of a telephone system. Wheeler was even a director of the Chicago-based Automatic Electric Company, which would supply the necessary telephone equipment.

Prior to the organization of the Illinois Telephone & Telegraph Company, Wheeler's only involvement in Chicago transportation was with the General Electric Railway Company, a small and rather obscure street railway enterprise. Incorporated on February 18, 1895, the GE was backed by a group which allegedly included several Chicago aldermen. The City Council subsequently granted the company potentially lucrative franchises to operate street railway service on a number of south side streets. One of Wheeler's associates in this endeavor was Edward J. Judd who would go on to help him organize the IT&T.

The General Electric's primary purpose was to acquire franchises for later resale to one of the city's several operating street railway companies at vastly inflated prices, turning a quick profit in the process. Their primary target was the well-established Chicago City Railway (CCR). Initially, the CCR was not particularly interested in what the GE had to offer and considered their overtures to be a form of legalized extortion. This was because if the CCR declined to buy the franchises, the General Electric would be in a position to operate a competing service. These suspicions prompted CCR to contest the validity of the franchises in court. The CCR lost several rounds to the General Electric and eventually bought most of the franchises in question for a reported $1 million.

In June 1898, it was reported that control of the tiny company had passed to "friends of the Chicago City Railway," effectively eliminating any competitive threat. This situation was not unique. A similar situation occurred in the gas distribution field when a group of aldermen organized the Ogden Gas Company.

While a tidy sum had been realized through the sale of the GE's franchises, Wheeler received little compensation – there was simply not enough money to grease the palms of all who those who had been involved in the scheme. In an effort to earn the fortune that had eluded him, Wheeler, along with Judd and several others, became active in the promotion of the Illinois Telephone & Telegraph Company.

It was not until Wheeler had gone on to help create the IT&T in 1898, that the General Electric ran its one and only car. The initial section to be placed in operation was located in Plymouth Court and extended a mere 800 feet south from Polk Street (the line eventually grew to about a mile in length). Service was provided by a battery-powered streetcar which, according to an article in the Chicago *Economist*, had come from the Englewood & Chicago Railway – a nearly as obscure street railway located on the city's far south side. Regular operation ceased in 1901, however trips were made on a sporadic basis until some time after 1912. Wheeler passed away on September 24, 1917.

at the World's Columbian Exposition which was held in Chicago in 1893. In a short period of time dial telephone technology had been perfected, rendering the original push-button technology obsolete. In 1901, the company was reorganized as the Automatic Electric Company, with Wheeler serving as its secretary-treasurer for a time.

On January 22, 1899, the IT&T's telephone franchise was finally approved by the City Council, only to be vetoed by Mayor Carter H. Harrison. In a lengthy letter to the Council dated January 30, Mayor Harrison detailed his objections to the franchise ordinance: the failure to specify a date when the telephone system had to be operational; allowing IT&T crews to open streets (including those that were newly paved); failure to explicitly give the city the right to use any poles erected by the company; and also that the 50-year term was too long (the mayor cited a similar situation in Indianapolis where a competing telephone company had been given only a 25-year franchise).

On February 6, the Committee on Gas, Oil and Electric Light sent to the full Council a revised ordinance that was designed to satisfy Harrison's objections. The matter was taken up at the Council's February 15 meeting. The final version was passed on February 20 and sent to the mayor.

The revamped franchise had a 30-year term and stipulated that all cables in the area bounded by Fullerton Avenue, Western Avenue, 22nd Street, Halsted Street, 55th Street and Lake Michigan be placed in underground conduits. Wires could still be brought above ground for short distances to reach subscribers.

It was also stipulated that a telephone exchange serving 2,000 customers be in service within five years. Harrison was still not entirely pleased, but allowed the ordinance to become law without his signature. It is interesting to note that the franchise did not specify the conduits' size, a loophole that the company would exploit to the greatest degree possible.

Although the franchise mentioned the use of tunnels to house wires, it was probably in the context of the already-existing LaSalle, Van Buren and Washington tunnels. These bores allowed cable cars belonging to two of the city's largest street railway companies to cross under the Chicago River to access the downtown area (construction of new tunnels to house telephone cables was not mentioned or even required). The *Street Railway Review* credited George W. Jackson with the decision to build tunnels instead of small-diameter conduits. According to that publication: "He (Jackson) found that the space below the paving was almost completely taken up by gas pipes, sewers and conduits for other companies, so it was decided that a tunnel system was the only practical solution of the problem at hand." Another justification given for building tunnels was so the workmen could easily unspool and install the telephone cabling, thereby speeding-up the installation process.

Bruce G. Moffat Collection

Joseph Harris in 1910.

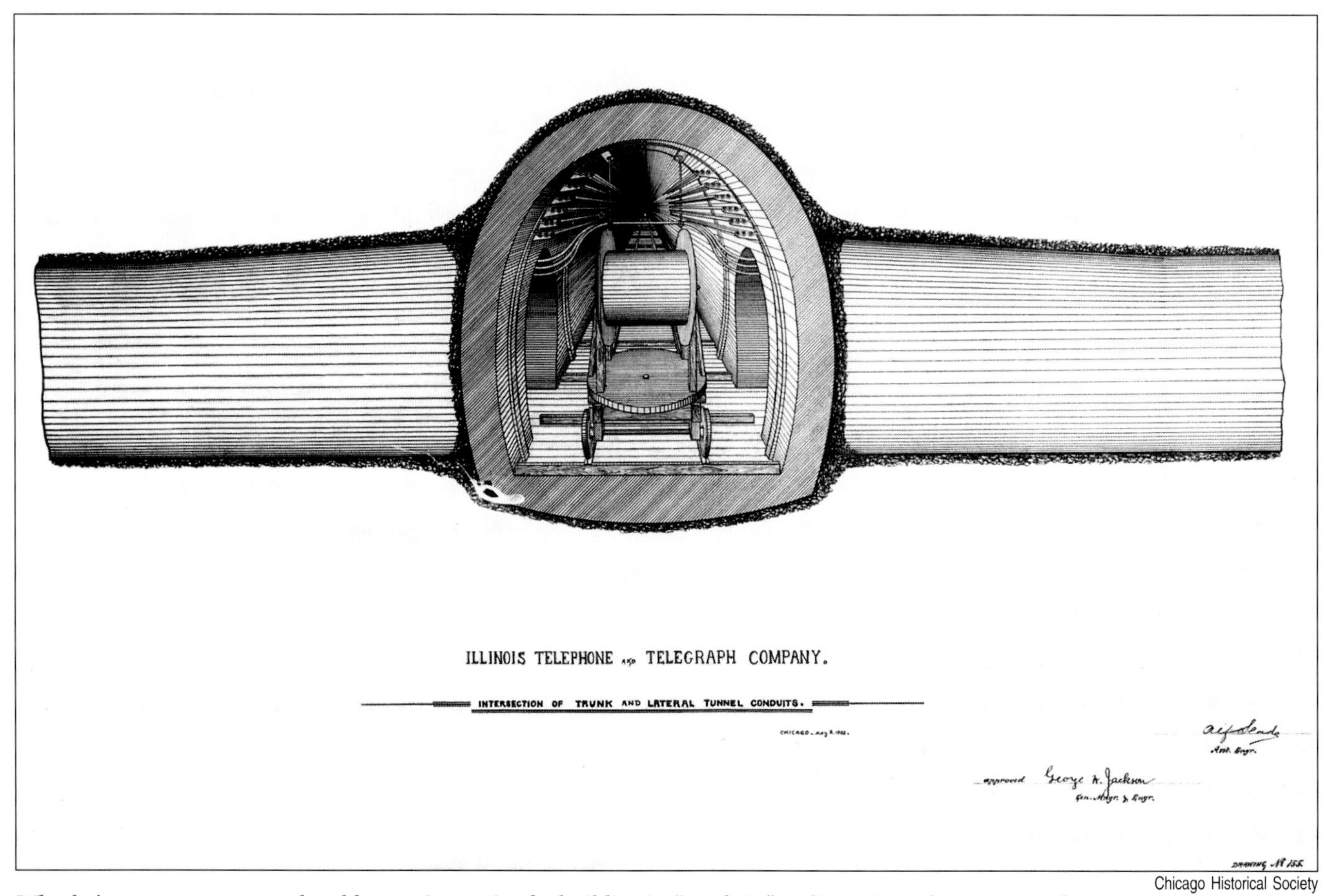

Chicago Historical Society

Wheeler's company came under aldermanic scrutiny for building its "conduits" to dimensions that were much more generous than the term normally implied. This 1902 engineering drawing was probably intended to bolster the company's assertion that their large size was needed so that large reels of cable could be easily unspooled and strung.

Bruce G. Moffat Collection

Mayor Carter H. Harrison in 1897.

At this point, Aldrich retired from the project while Wheeler chose to remain and became the company's president. One of his early acts was to organize a construction subsidiary to handle the actual building of the system. This was a common practice designed to insulate the operating company from risks associated with unforeseen cost overruns or capitalization problems.

The Illinois Telephone Construction Company was incorporated on August 14, 1899, by Albert G. Wheeler, Edward J. Judd and Louis Behare. Of the 5,000 shares of stock issued, Wheeler held all but two. Speculation was rampant that the Strowger company intended to purchase the IT&T's franchise and build and operate the system themselves. This did not happen. Instead, the Strowger company largely confined its efforts to promoting the development of an "independent" telephone network and supplying the constituent companies with all of the necessary telephone equipment.

Hopes were high for Strowger's technology. A *Tribune* writer commented:

> The practicability of the new system, which will do without the services of the "hello girl" for one thing, has been so thoroughly proved by extensive operation, it is said, as to leave no doubt as to its excellence. It is asserted the new system not only will possess the virtue of securing absolute secrecy between the two at different ends of the wire, but also can be operated much more expeditiously, and at a much less expense than the present system. It is calculated that on an exchange of 10,000 subscribers a savings of over $500 a day can be made.
>
> By this system all a person has to do to call any other subscriber is to pull down a small lever to the numbers indicated on the dial... The exchange will be absolutely automatic.

ANNOUNCEMENT TO CHICAGO TELEPHONE SUBSCRIBERS

LOOK IN YOUR MORNING'S MAIL

FOR A PROPOSITION FROM

The New Telephone Company.

PROMPT—PRIVATE—PERFECT SERVICE.

REBATE TO FIRST 10,000 SUBSCRIBERS.

30,000 Telephones to be installed at Earliest Possible Date.

PRICES.		More Subscribers.	EQUIPMENT.
Business Telephones,	$85.00	Better Service.	Long - Distance Phones; Metallic Circuits;
Residence Telephones,	$50.00	Lower Charges.	No party lines; No limit to calls.

Section 9 of the ordinance under which this company will operate: "It shall be expressly the condition of this grant that if the Illinois Telephone and Telegraph Company, or any of it's successors or assigns, shall either sell out to or enter into any agreement with any existing telephone company, or any of it's successors or assigns, doing business in the City of Chicago, which agreement would tend to make competition inoperative, this ordinance shall become null and void, and the plant of said company, together with the conduits, wires and poles then in the streets belonging to said company, shal be forfeited to the city."

Any telephone user who fails to receive above proposition and all others who have been deprived of telephone serice on account of high rates charged in the past are invited to send to this office for information.

Illinois Telephone and Telegraph Company.

721 - 723 Rookery Building

ALBERT G. WHEELER, President **L. H. ALLEN, Contracting Agent**

Bruce G. Moffat Collection

On August 4, 1899, the Illinois Telephone & Telegraph Company placed this advertisement in Chicago's morning newspapers soliciting subscribers for its as-yet-unbuilt automatic telephone system.

Unfortunately, what the writer didn't say was that there was no planned interconnection with the lines of the other local telephone company, thereby limiting the new system's reach. In all probability, the Chicago Telephone Company would have actively opposed any such initiative.

Strowger's Joseph Harris was so confident in his product that he contemplated issuing a challenge to anyone who could prove that the automatic system was inadequate for use on a large-scale basis. The June 13, 1899 *Tribune* reported:

> J. Harris, secretary of the Strowger company, said: "Within a short time we intend issuing a challenge for $5,000 or $10,000, open to any one who claims that our automatic telephone connecting apparatus will not do the work we say it will. Several persons have said this automatic appliance must be a failure in a large exchange. We are willing to leave it to experts. We will put the money up in a bank, and will give the sum gained to charitable institutions."

Among the doubters was Chicago Telephone Company President J. M. Clark, who expressed his confidence in the Bell System's manual technology when he stated: "I do not see how this automatic device...can possibly be a success at large telephone stations. I have no doubt it works well on a small scale, as have experiments made in our offices. This year we will expend some $1,500,000 on our buildings, apparatus, and laying underground conduits. No new company will make any change in our prices or have any effect on our service. We lately took on 100 additional women telephone operators, which brings the total up to 700, and it will not be long before the number will be increased to 1,000."

On August 4, 1899, Chicagoans opened their newspapers to find a paid "Announcement to Chicago Telephone Subscribers." This eye-catching advertisement alerted Chicago Telephone Company subscribers to check their morning's mail (mail was delivered twice daily in Chicago at the time) for a "proposition from The New Telephone Company." The primary purpose of this unusual mass mailing was to call attention to the superior service that prospective subscribers could expect from the Illinois Telephone and Telegraph Company. The advertisement seemed to be a bit premature since the company estimated it would be about six weeks until the actual groundbreaking (no estimate was given on when the first exchange would be operational).

Simultaneously with the advertisement's publication, a team of surveyors was sent out to map various downtown streets where the company intended to lay cables. Actual construction began shortly thereafter at a most unusual location.

Construction

The first construction shaft was sunk in the rear basement of a building at 170 (now 165 West) Madison Street. The building's main floor housed the Powers & O'Brien Saloon. This downtown watering hole was operated by Alderman Johnny Powers and was located less than two blocks from City Hall.

The basement was completely filled with all of the construction equipment and materials necessary to conduct an ambitious tunneling effort. Sand, gravel and cement were fed into a concrete mixer by means of a traveling belt. The resulting mixture was loaded into small construction cars and lowered into the shaft by a steam-powered elevator.

At the base of the construction shaft, thirty feet below street level, a small-diameter construction "conduit" (tunnel) measuring about four feet in width and height was built into the alley where it expanded to four feet wide and seven feet high. The tunneling crews then burrowed eastward to the center of LaSalle Street where a larger "conduit" measuring 6'9" wide and 7'6" high was dug under that street from Madison Street to Monroe Street, a distance of one block. Completed sometime in early 1900, this segment became the standard-size bore for most of the projected system.

Construction crews were organized on a three-shift basis so that work could continue around the clock. The two night shifts performed the actual tunneling operations, cutting through the clay with hand-held knives, while the day shift built the concrete walls. Tunneling operations were generally carried out under ten pounds of air pressure to minimize the possibility of cave-ins, permitting daily progress of 12 to 16 feet at each construction heading.

To build the walls, workmen first assembled wooden forms. Concrete was then hand packed into the cavity behind the forms and left to dry. The forms were then removed and an additional layer of concrete was applied to give the walls a smooth appearance. In later years, this finish layer was often omitted,

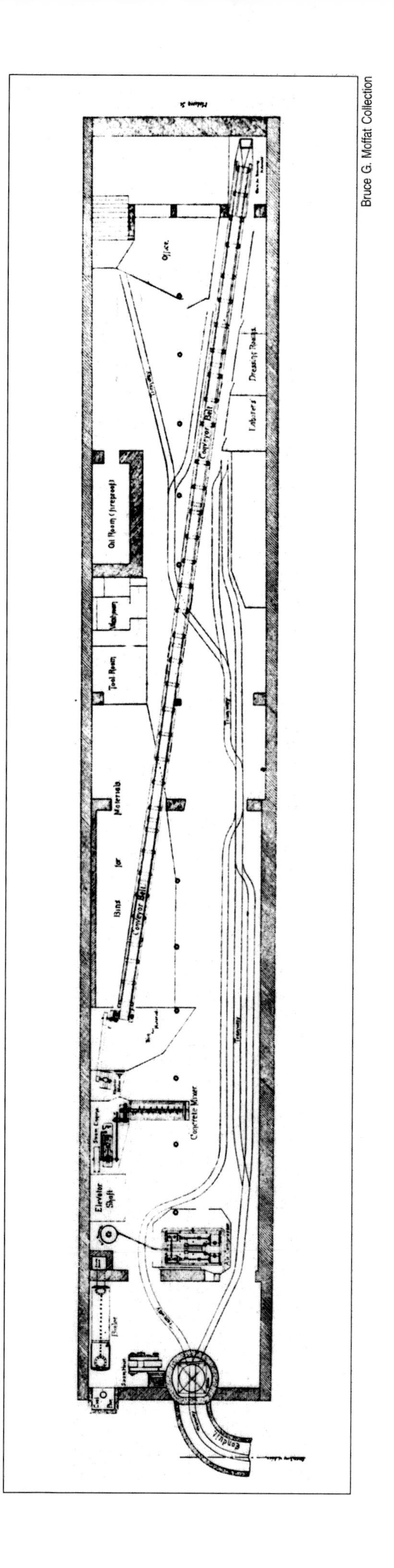

Bruce G. Moffat Collection

This floor plan of the basement beneath Powers & O'Brien's Saloon illustrates the cramped layout of the company's first "shaft." 14" gauge track was installed for handling cars of spoil and construction supplies. The elevator connection to the alley tunnel is at the bottom of the illustration.

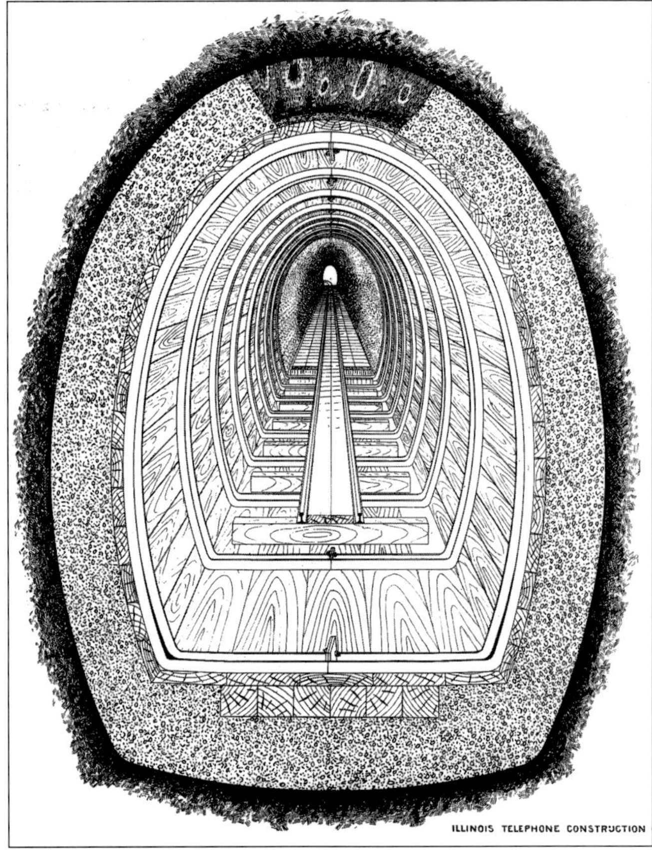

Bruce G. Moffat Collection

Engineering drawing showing the wood framing used to form the concrete walls. Also shown is the temporary 14" gauge construction track. Once the concrete had set, the wooden forms were removed and a finish coat of cement was applied to the walls.

Bruce G. Moffat Collection

Photo of actual construction. Note the gouge marks in the surrounding clay that resulted from the miners having to carve out this heavy substance using special knives.

allowing the grain marks from the wooden forms to remain visible. The completed walls averaged twelve inches in thickness, and have remained largely crack-free even though they lacked any steel reinforcement and despite the construction of new buildings nearby.

Initially, the construction spoil (excavated clay and dirt) was raised into the saloon's basement and brought up to the alley using a small elevator and loaded into waiting wagons. The wagons frequently blocked the narrow alley, resulting in complaints from neighboring businesses. A more serious complaint arose in November, 1899, when the owners of a building abutting the alley near the saloon complained to the city that their alley wall had developed cracks "from the settling process." Needless to say, they were concerned that the tunneling might be undermining their building and asked the city to investigate. The results of the investigation were apparently inconclusive. This was but the first of many complaints alleging structural damage caused by the work.

Allegations of Forgery

Settling problems notwithstanding, some aldermen were still suspicious about the company's plans and whether the construction work was being managed in accordance with the franchise. The aldermen directed the public works commissioner to conduct an investigation. On February 5, 1900, the commissioner submitted a generally favorable report. Alderman Novak moved that the report be approved and placed on file.

The commissioner's report was quickly forgotten. Then, in September 1904, the company's former attorney, Edward J. Judd, made a spectacular allegation. Judd alleged that Wheeler had conspired with then-City Clerk William Loeffler and former Deputy City Clerk Edward Ehrhorn to falsify an entry in the City Council's Journal of Proceedings for February 5, 1900.

The backdrop for this tempest was an apparently long-running feud between Judd and his former employer in this matter. At one point, news articles mentioned that Judd even succeeded in having Wheeler and Harris arrested but then declined to press charges. Judd had earlier appeared before several judges to present his allegations, but they failed to find any merit in his claims. Wheeler and Harris charged that Judd was trying to blackmail them because he threatened to release potentially damaging information about the company's activities.

All rhetoric aside, the basic controversy centered around whether the commissioner's report was

Bruce G. Moffat Collection

A group of workmen pose at the intersection of the Madison and Monroe bores. Note the temporary turntable used for routing the construction cars.

"approved and placed on file" as stated in the proceedings or, as Judd alleged, that the entry should have simply stated "placed on file." The latter designation would not have conferred official concurrence in the commissioner's findings or other formal endorsement by the lawmakers.

Judd's credibility was quickly called into question when reports suddenly appeared in the press announcing that he was being arraigned for embezzlement. The complaint was brought not by the IT&T but by a Miss Mary C. Overlook who alleged that he had diverted $69 that she had given to him to pay her medical expenses.

Unsuccessful in his initial efforts to have a grand jury convened, Judd pressed ahead with a lawsuit of his own charging that the defendants had conspired to forge official records. The trial commenced at the end of November 1904. Three witnesses testified for the defense before Circuit Court Judge Everett. Council Reading Clerk Wyatt McGaffey claimed to know nothing about the alleged "correction." He also testified that the only rubber stamp in the office said "placed on file," so he used it.

His testimony was followed by that of James W. Higgins who had supervised the printing of the Council's proceedings. He stated that after 150 copies had been printed, Ehrhorn called him and directed his attention to an error. Ehrhorn told him that after the report on the IT&T's activities the record should state "approved and placed on file" instead of "placed on file." (Judd had earlier testified that he was present at that session and that Novak had made no such motion.) According to the *Tribune*, Higgins admitted that the item had been marked for correction on the proof sheet but that it had been overlooked when the final typesetting was done.

Alderman's Novak's testimony could only be described as frank. The alderman freely admitted that he had not read the commissioner's report before moving for its approval. Novak's testimony supported the defense's claim that the discrepancy was nothing more than a printer's error. When asked why he moved approval of a report that he had not read, he simply replied "I was asked."

That was all the judge needed to hear and promptly ruled that there was no evidence to support the contention that Wheeler and the other defendants had conspired to alter the record and ordered the defendants released.

The defendants' troubles were hardly over, however. Not only was Judd still determined to bring down Wheeler, but the Chicago Federation of Labor (CFL) was ready to enter the fray. On December 24, 1904, the *American* reported that the labor organization had been conducting a three-month investigation of the company's activities and the roles played by Wheeler, Loeffler and Ehrhorn. The CFL was not happy that the company was making use of non-union construction forces and that the completed bores had the potential of putting many dues-paying teamsters out of business. The CFL's special investigation committee was in close contact with Judd and State's Attorney Healey as they worked to build a case.

In early January 1905, a grand jury investigation was launched to look into Judd's allegations and to evaluate the evidence gathered by the CFL. Among the new allegations was that Alderman Novak had lied when he testified before Judge Everett. It was also alleged that Ehrhorn and Loeffler had received tunnel securities in exchange for assisting the tunnel's promoters and that some aldermen had been bribed to gain passage of the original franchise grant in 1899. The principal witness was Judd. Perhaps his most startling revelation was that on February 21, 1899, the day after the franchise's approval, he had personally delivered $110,000 in company funds to the Powers & O'Brien Saloon where it was distributed among 21 waiting aldermen who had voted in favor of the grant.

The defendants were quick to assert their innocence. The *Tribune* quoted Novak as saying:

> Of course I deny my guilt. I do not ask the public to believe me, but to weigh carefully the character, record, motives, and chimerical stories of the principal accuser. He [Judd] is in striking contrast to the men whom he accuses. They have at least at times shown that they were capable of filling responsible positions with credit.

The Chicago Federation of Labor's report contained this bombshell:

> We find that Perry Hull, a former well-known politician and E. Judd, formerly connected with the county attorney's office, conceived the idea of a tunnel system for the city. To secure the necessary franchise it was important to do two things – secure the friendly cooperation of corrupt alderman and conceal from the public their real purpose. Accordingly they drew up a franchise permitting them to build a conduit for telephone and telegraph purposes. Evidence furnished is that they got the aid of council members on a promise of $110,000 to be divided among them upon passage of the ordinance, and they deceived the honest members of the council and the public at large by claiming that their only purpose was to build a competing telegraph line and thereby destroy the monopoly of communication.

The report went on to assert that when the St. Louis financiers backed out of the project it was Judd who recruited Wheeler to find new financing and carry out the project. Unfortunately there survives no independent confirmation that Judd had the original inspiration for this project.

On February 4, 1905, the grand jury indicted Wheeler for forgery, Novak for forgery and perjury, Ehrhorn for forgery, and Higgins for forgery and perjury. In its report to prosecutors, the grand jury said: "Our investigation of cases involving vast interests has shown us criminal carelessness on the part of the city officials in issuing permits granting greater powers than those contemplated in ordinances."

On June 19, 1905, the case against Wheeler and his accomplices was heard before Judge Arthur Chetlain who promptly quashed the indictments. In the furor that followed, the Chicago Bar Association censured Chetlain, but miraculously he escaped disbarment for lack of sufficient evidence. The Bar Association did note the existence of a promissory note, signed by Chetlain, which was held by persons interested in having the indictments quashed. Certainly, the appearance of a conflict of interest was there for all to see.

This marked the end of Judd's efforts against the company and those connected with the granting of the 1899 franchise. However in a curious footnote, the private Citizens Association attempted to hold Ehrhorn accountable for the alleged falsification by pressing charges before the city's civil service commission. Barely had the group finalized its charges than the *Post* announced the loss of a critical piece of evidence:

> Through the alleged carelessness of Reading Clerk Wyatt McGaffey, the only copy owned by the city clerk of the original council proceeding containing the report of the Illinois Tunnel Company ordinance has been destroyed, and if charges are filed against Deputy City Clerk Edward H. Ehrhorn for altering the council record, the commission will have to seek its evidence elsewhere.
>
> McGaffey had charge of the council proceedings, of which copies are extremely scarce, and as soon as the trial in the Criminal Court was ended he says he threw the document into the waste basket with other discarded papers, believing it to be of no more use.

This apparently ended all further prosecutorial efforts.

2 Building a Railroad

By March 1900, four additional construction shafts were in the process of being built. Unlike the original installation in the basement of the saloon, the new shafts were capped with head houses built on Loop sidewalks. The head houses were built of heavy timbers and sheathed in corrugated galvanized iron and measured six feet square and 23 feet high. The shafts themselves were circular and contained a construction lift (elevator) to carry supplies and construction spoil. To supply the 10 pounds air pressure needed at the work sites to prevent cave-ins, space was leased in building basements adjacent to the shafts to house air compressors, steam powered hoisting machinery and a small dynamo to power the compressor and the tunnel lighting. The temperature at the tunnel headings under air pressure was generally in the 75-85 degree range. Air locks were used to separate the work headings from the completed tunnels. A small telephone system was also installed to transmit progress reports from the construction shafts and the chief engineer's office then located in the Auditorium Hotel Building.

According to the *Western Electrician*, the company's tunneling ambitions extended far beyond the city's downtown:

> About 11 miles of tunnel is to be built in the business district. It is expected that two branches will be extended under the Chicago River, at a depth of 60 feet, to connect with the North and West Sides of the city. The farthest points which the company proposes to reach by means of its tunnels are North avenue, by way of Franklin street, on the north, Western avenue, by way of Madison street on the west, and Sixty-seventh street, by way of State Street, on the south.

Although the system never reached any of these far-flung points, the company managed to build tunnels spanning the river at not two, but 13 locations.

During the summer of 1900, all construction activity was suspended. The construction company's engineers claimed that inaccuracies had been discovered in the city's maps and that additional surveying was required before work could be resumed.

Chicago Historical Society (ICHi-23270)

A workman feeds telephone cable into a manhole at the intersection of La Salle and Madison.

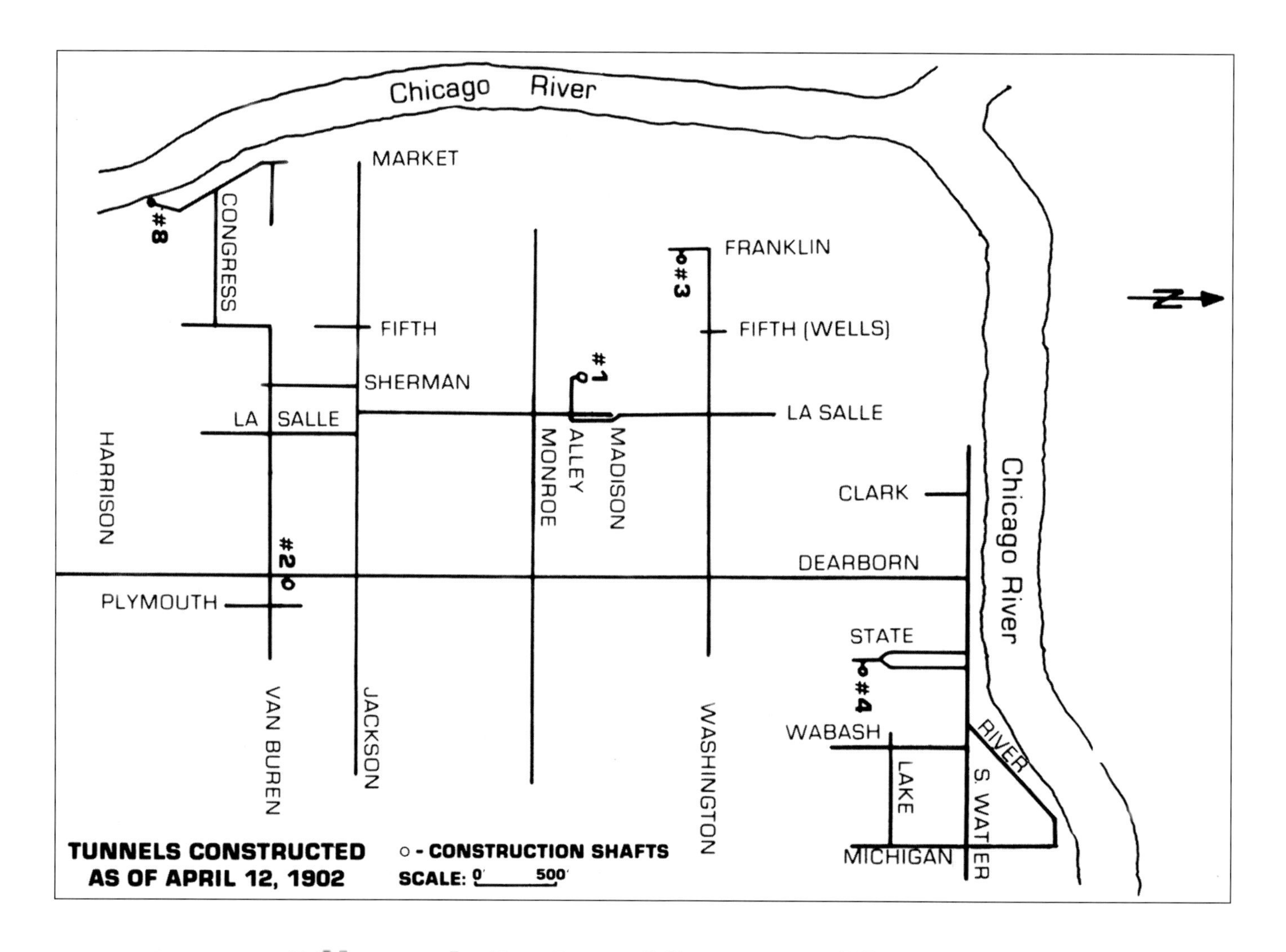

A Change in Grade and Corporate Direction

While construction may have ceased, the activities of the promoters definitely had not. Without any public notice, they quietly moved to increase the corporation's powers to include activities unrelated to telephone service. At the company's regular annual stockholders meeting held on May 7, 1900, President Wheeler proposed that the company's articles of incorporation be broadened so that it could "...distribute, convey and deliver, telephone messages, telegraph messages, packages, mails, newspapers and general merchandise, and to convey, distribute, deliver and furnish power, heat and light by steam, water, air, electricity or otherwise, but not by railroad; also in connection therewith to manufacture, buy, sell and deal in all kinds of instruments, appliances, apparatus, licenses, patents, patent rights, and to take and to hold property necessary or useful in carrying out such object; also to acquire, construct, build, buy, sell, lease to or from others, and to operate such plant or plants as may be necessary or useful in carrying out such object." Predictably, the proposition was approved unanimously.

In essence, this change reflected the promoters' intentions to use the tunnels not only for railroad purposes, but also to install pipes and conduits for the commercial transmission of steam heat and possibly electricity. From this point forward, emphasis was placed on building a railway, with telephone operations assuming a strictly secondary role.

With surveying complete, construction was resumed in September 1901, but at a greater depth than before. The change in tunneling depth from 30 to 40 feet was apparently ordered so that tunneling could be done largely in the blue clay stratum that underlies most of the downtown district. This simplified digging, lessened the potential for cave-ins, and minimized the likelihood of disrupting the orderly settlement of adjacent buildings. All tunnels were subsequently constructed at this lower depth except near the Chicago River where they ran even deeper in order to pass beneath the waterway at a safe level.

The original block-long segment under LaSalle Street was retained at its original depth. However in order to connect it with the new lower tunnels, the Monroe Street bore rose up the necessary ten feet to meet the "upper" LaSalle tunnel. Similarly, the extension under LaSalle south from Monroe quickly descended to the 40-foot level. The other shafts were quickly extended to the new level and several new mining crews were put to work.

Western Electrician

Shaft #2 was located on the east side of Dearborn just north of Van Buren. Note the arc street light in this early 1900 view.

In order to provide construction forces access from the original construction shaft to the work sites north of the intersection of Madison and LaSalle, a special connecting tunnel was built. Called a "monkey drift," this passageway started opposite the point where the original construction tunnel met the LaSalle tunnel and extended northward, parallel to the LaSalle tunnel, but descended rapidly to the new grade. Because this bore was built to a slightly smaller profile than the regular main tunnels and the steeper gradient, track was never installed. All of this construction activity still left the north end of the original LaSalle tunnel without any connection to the other tunnels being built. Since freight connections were never made to buildings in this block, this dead-end tunnel could be considered a waste of construction effort and was relegated to the storage of surplus freight cars.

Western Electrician

An adjacent building basement was used to house necessary support equipment including a small electric dynamo and an air compressor.

Goodman Equipment Corporation

An artist's conception of what a typical intersection would look like once the railroad and telephone systems became operational. The entire ceiling area was to be occupied by telephone cables while Morgan-built locomotives, drawing current from a center cog rail, would handle the freight traffic. The size of the tunnel is greatly exaggerated.

Spoil Removal

From the start of construction in 1899, spoil removal was a major traffic source.

During the first year of construction, excavated material was hauled up to the surface through the original shaft in the basement of the Powers & O'Brien Saloon and transported by wagon to low-lying areas in need of fill. As construction moved farther away from the saloon and other construction shafts were opened, it became necessary to use special mule-hauled cars running on temporary 14" gauge track to carry the spoil to the nearest shaft for removal.

In 1902, a special disposal station was opened along the east bank of the Chicago River where the Eisenhower Expressway now crosses. Officially known as Shaft #8, it was designed to handle large quantities of spoil rapidly. The station was equipped with a link-belt incline that brought the construction cars to the surface where their contents were transferred to scows for final dumping into Lake Michigan. This station was closed by 1908. (This was also the location of the Illinois Telephone Construction Company's shop and stable.)

In 1904, a commercial disposal station was opened along the west bank of the river between Washington and Madison Streets. Here, the loaded cars were lifted out of the tunnel using a hoist and their contents dumped into scows. This facility was owned by George W. Jackson who also served as general manager and chief engineer of both the Illinois Telephone Construction Company and the Illinois Tunnel Company. Loads brought here by wagon were from non-tunnel sources such as building construction sites. This station remained in service for several years.

By 1909, the tunnel network was largely complete, with only a few minor extensions and building connections added in later years. Among these were the Field Museum (1921), Builders and Pittsfield Buildings (1927) and the Merchandise Mart (1932). The last connection made was with the Prudential Building in 1954. The Prudential's connection was never used, but was installed as a contingency in the event that a future scarcity of gas and oil fuel would require the building to convert its boilers to burn coal.

Right: *A teamster receives a load of clay spoil at Shaft #7 which was located on Eldridge (now 9th) just east of State. Never averse to self-promotion, Jackson made sure his name and phone number was conspicuously displayed.*

Bruce G. Moffat Collection

George W. Jackson owned and operated this disposal station located on the west side of the Chicago River between Washington and Madison. It was demolished in the 1920's to make way for construction of the Chicago Daily News Building.

Bruce G. Moffat Collection

Construction Shaft #8

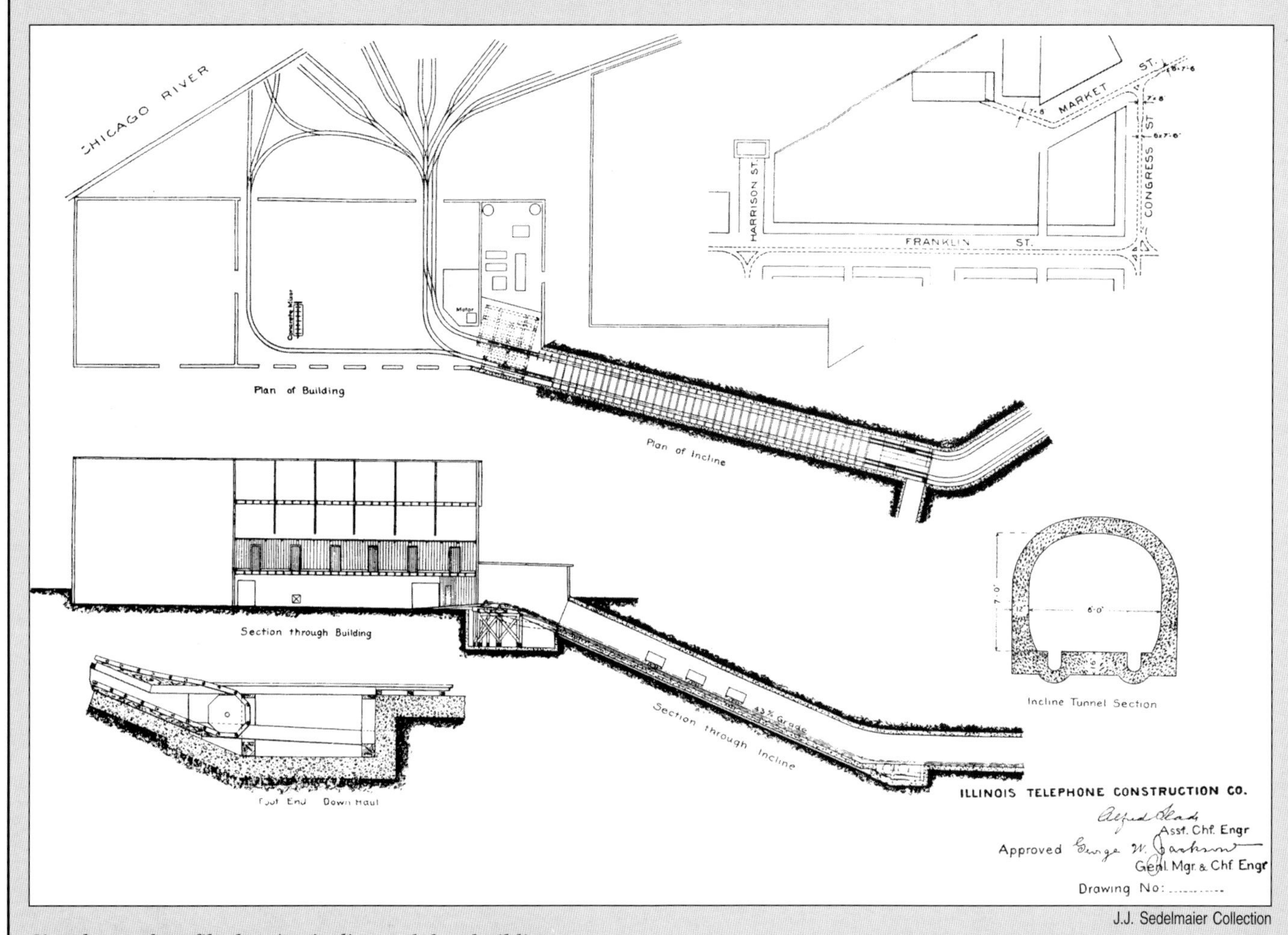

J.J. Sedelmaier Collection

Site plan and profile showing incline and shop building.

Shaft #8 was unlike any of the company's other construction sites. Located at about where today's Eisenhower Expressway passes over the east bank of the Chicago River, this facility eschewed the usual construction elevator arrangement. Instead, two Link Belt conveyors were installed to raise the 14" gauge spoil cars to the surface (the wide space between the two tracks was used as a walkway). At surface level the tracks passed though a four story building housing the company's machine and blacksmith shops before ending at a small dock where the spoil was dumped into a waiting scow. Also on site was a stable that housed the large number of mules required to pull the cars through the tunnels.

Chicago Historical Society (ICHi-23277)

The "mule camp" was a crowded place indeed.

A look down the incline showing the two tracks alongside either wall.

Bruce G. Moffat Collection

Larry Best Collection

The wood and blacksmith shops.

Construction Shaft #8

Chicago Historical Society (ICHi-23281)

Gangplanks were used to position the spoil cars for unloading.

Chicago Historical Society (ICHi-23291)

The compact dock area.

Equipping the Tunnels for Railroad Operations

Following the decision to engage in freight service, installation of rail and wire in the completed tunnels was begun. In those bores still under construction, installation of rail and overhead trolley wire immediately followed completion of the concrete work. In some cases, this resulted in track extending a few feet past a switch only to end at a blank wall because further construction had been called off.

Rail was delivered into the tunnels through chutes that had been installed at a number of locations. These chutes consisted of long pipes driven into the ground at an angle. The rail, which weighed 56 pounds per yard, was simply placed into the pipe and allowed to slide down into the tunnel where it was picked up and transported to the work site. For the most part the rail was secured to cast iron rail chairs that were embedded in the concrete tunnel floor. In later years some track was laid using 60-pound rail resting on wooden ties. In either case, the space between the rails was covered with a layer of concrete, completely covering the rail chairs (or ties). In some locations the concrete center strip was brought level with the top of the rail.

The overhead trolley wire, feeder cables and related fittings were of the type normally used on mine railways. Copper 4/0 (0000) grooved trolley wire was used in most locations. Electric switches activated by trolley contactors were employed at junctions. Switches leading to sidings were of the manual type often associated with street railway lines. These required the train's operator to use a switch iron or similar device to align the switch points for the desired route. If necessary, wooden blocks could used to hold the points in place. In later years some switches were modified so that they could be operated by having the operator pull on a long chain which was connected to the switch points by means of an ingenious system of pulleys.

Because of the possibility of cave-ins and settlement problems caused by tunnel construction, the city took an active interest in the work. To insure that only quality materials were used, and that proper engineering and construction practices were being adhered to, the City Council directed the public works commissioner to undertake yet another investigation. William R. Northway, an assistant engineer in the department's Bureau of Engineering made an extensive inspection of the work sites and authored the following report to his boss on December 21, 1901. His report was then referred to the City Council and read at their meeting of January 13, 1902:

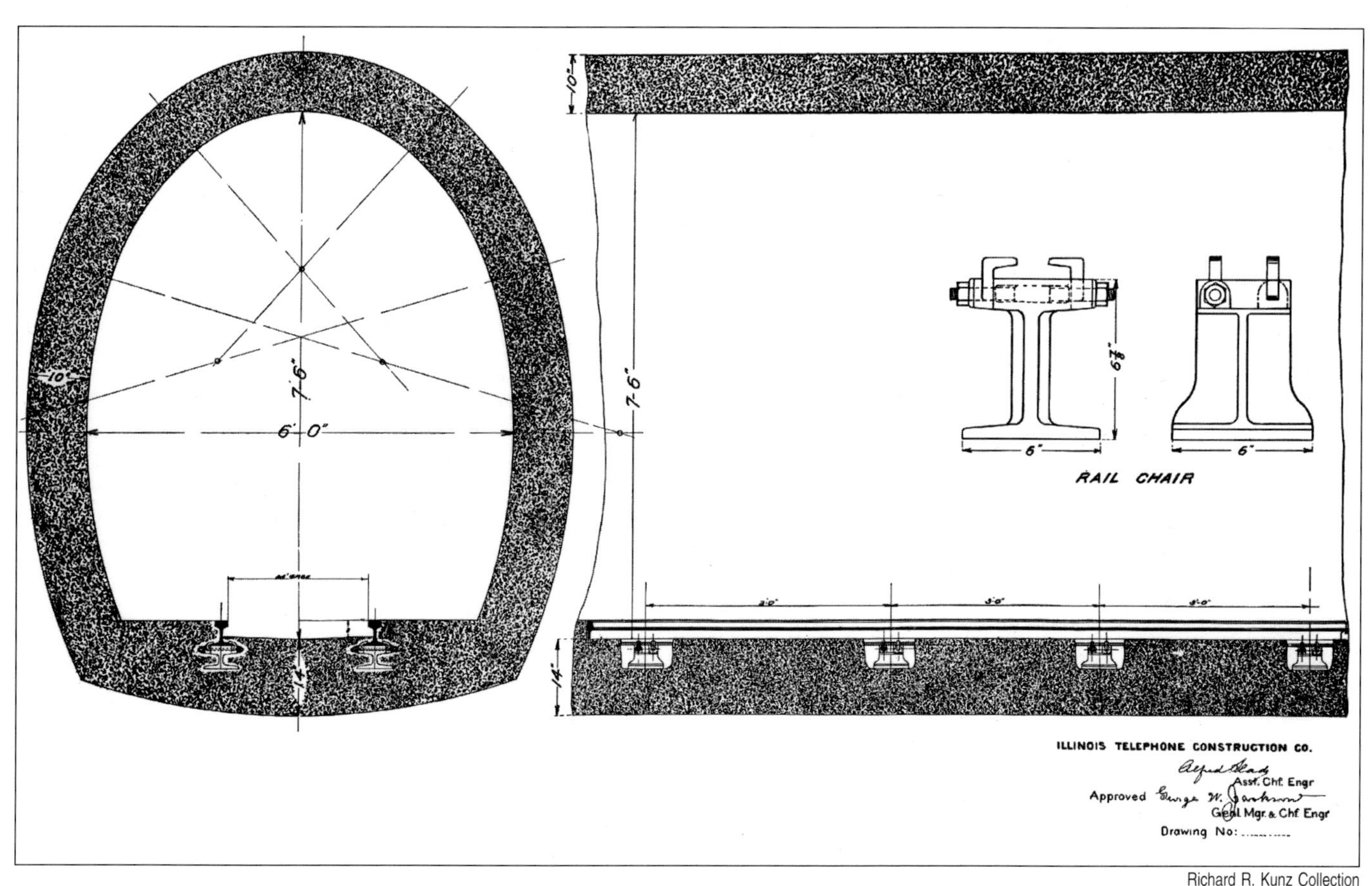

Richard R. Kunz Collection

Drawing showing finished tunnel profile and the cast iron chairs used to support the running rails.

Mr. John Erickson, City Engineer:

Dear Sir:

In compliance with your commu-nication dated the 19th inst., and the order of the Council accompanying, which is herewith returned, I have to report as follows:

The material used in the tunnel of the Illinois Telephone and Telegraph Company up to about three weeks ago was concrete made of Atlas cement one (1) part, torpedo sand two (2) parts, and crushed stone five (5) parts. Since that time the material has been clean gravel in the proportion of one (1) part Atlas cement to four (4) or five (5) of gravel, as the proportion of sand increased or decreased in the gravel.

I have daily examined the gravel delivered and directed more cement to be used when the proportion of sand increased, which has always been done. The gravel used is of the best quality and I have never found any soil left on the hand in trying it.

The amount of concrete used is about one cubic yard per foot of tunnel; sometimes where loose mining has been done the amount is materially increased.

In a number of cases where it has been necessary to cut out concrete I have found the work to be well done and the concrete hard and solid, and I have not found any material difference between the concrete made with crushed limestone or gravel; if any difference, it seems to be in favor of the gravel.

The company has two inspectors of its own who visit each shaft and all the drifts where work is progressing twice each day and look after the proper thickness of concrete, as well as the proper proportions of cement and gravel or other material. If by quantity is meant the number of feet constructed, 7,700 will be a very close approximation.

Yours truly,

/s/ Wm. R. Northway
Assistant Engineer

Chicago Historical Society

The temporary street signs on the wall indicate that this is the intersection of Van Buren and Dearborn. The 14" gauge construction trackage and temporary overhead lighting are still in place. Work crews would soon return to this location to install the permanent 24" gauge track.

Chicago Historical Society (ICHi-23280)

The bustling activity and intense traffic congestion at the Wabash Railroad's freight house on Federal Street was typical for the city's numerous freight houses in 1904. The Tunnel Company hoped to lure much of this traffic away from the teamsters.

Following the decision to use the tunnels for freight movements, IT&T officials continued to let city officials believe that its sole goal was to construct and operate a telephone system. The company's construction activities aroused the suspicions of some of the aldermen, however. The aldermen created a special committee to determine just what the company's promoters were really up to. The committee, comprised of three aldermen, the public works commissioner and the city electrician, toured the rapidly expanding tunnel network and held meetings with company officials. The committee's findings were presented at the City Council's meeting on April 21, 1902:

> We beg to report that we find the smaller conduit is being built according to the size given (about 6 feet by 7 feet 6 inches) in their original permits, under the streets shown in the accompanying plat.
>
> The work is done in a very satisfactory manner and acceptable to the Commissioner of Public Works, his inspectors, and also to your Committee.
>
> Your Committee also finds that the Illinois Telephone and Telegraph Company has commenced operations at two different points on a much larger conduit than they had shown to the Commissioner of Public Works and City Engineer when they reported to this Council on February 5, 1900, and the Commissioner of Public Works had refused to allow work on these larger conduits to continue.
>
> The size of his large conduit is 12 ft. 9 in. by 14 ft.
>
> Your Committee does not feel justified in recommending to the Council that permission be given to authorize the continuance of work on this greatly increased size of conduit without a definite conclusion being arrived at by the City Council as to what size of conduit the Commissioner of Public Works issues a permit for.
>
> The evidence submitted to your committee

by expert engineers as to the size of the conduits necessary for this company to successfully carry on its business was of a conflicting character; but in the judgment of your Committee a tunnel of 12 ft. 9 in. by 14 ft. is not required for conducting the telephone and telegraph business outlined by the company to your Committee.

Following the "discovery" of these fairly short (generally 100-200 feet) "conduits" (the company claimed these were to be used as connection points for trunk telephone cables), a very serious controversy erupted over whether the company was making a "land grab" at the public's expense.

It was widely speculated that the company planned to operate an underground standard gauge railway. This line of thinking gained some credence when an article appeared in the March 8, 1902, issue of *Electrical World and Engineer* which included this telling passage: "The main or trunk tunnels are 14 ft. high and 12 ft. wide. The lateral or branch tunnels are 8 ft. high and 6 ft. wide. This gives sufficient room to not only hold all the telephone wires of the company and underground wires of other companies, but to provide for the operation of cars through the tunnel for hauling mails and handling freight."

Eventually, word of the company's real motive leaked out. Mayor Harrison reacted by ordering an immediate halt to further construction on April 27, 1902, on the grounds that use of the tunnels for transportation purposes would be contrary to the terms and conditions of the franchise, which did not address the handling of freight. The press was rife with articles about the company's "land grab" and how they were cheating the city and its citizens out of its rightful compensation. Some even advocated municipal ownership of the network.

The day prior to Mayor Harrison's action another inspection party toured the system – but for a very different reason, as reported in the *Tribune* of April 27:

From Harrison Street and the river to South Water Street and Wabash Avenue, 40 aldermen – over half the City Council – tramped and crawled through the downtown district, 40 feet below the surface of the streets yesterday afternoon. The biggest investigating body which ever represented the Council was trying to discover what effect the tunnel of the Illinois Telephone and Telegraph Company will have on the proposed city subway.

Almost the entire membership of the Local Transportation Committee, now considering plans for the readjustment of the traction lines, went through the bores. At a committee meeting held before the investigation was begun, George W. Knox, a civil engineer, outlined plans for a six-track subway. Then the aldermen went down into the tunnels to see what is left of the streets for the street railway lines…

Serious interference with the subway construction was predicted by members of the investigating committee, but officials of the company claimed that the subway could go directly on top of the telephone tunnel.

All plans for this streetcar subway proposal were subsequently dropped in favor of more ambitious and comprehensive Loop subway proposals that would encompass both street railway and rapid transit lines. In any event, the city's first subway (for use by rapid transit trains) would not be opened until 1943.

An Amended Ordinance

Eventually, an accommodation was reached between the IT&T and the city, and on July 6, 1903, an amended franchise ordinance, drafted jointly by the Judiciary and Local Transportation committees, was referred to the full Council for approval. The ordinance provided for the legalization of the existing tunnels and would allow the IT&T to construct additional tunnels and the right to transport freight, mail and other merchandise through them.

On July 15, the draft ordinance was brought up for debate at which time eight aldermen presented no less than 21 amendments. It was then approved by a vote of 64 to 2.

The franchise that emerged from this process authorized the construction and operation of a 50-mile system within ten years or the forfeiture of all rights and privileges granted along with all plant and equipment. The ordinance also stipulated that, upon the expiration of the franchise in 1929, ownership of all tunnels not located under private property would revert to the city. However the rails, conduits and all other equipment housed in the tunnels would remain under private ownership. This provision would require the city to purchase said equipment for a negotiated amount should it wish to take over and operate the system.

The city also reserved the right to order any tunnel to be removed, altered or relocated if necessary to accommodate the construction of water tunnels or passenger subways. The ordinance also limited the size of most tunnels to 6'9" wide and 7'6" high; however tunnels located under certain designated streets could be as large as 12'9" wide by 14'0" high.

The relocation clause was referenced in a later transportation plan calling for the construction of subways for use by both streetcars and rapid transit trains to be built on top of the tunnels. That report, issued by the Chicago Traction and Subway Commission in 1916, proposed the building of a shallow subway under State Street for use by rapid transit trains and a pair of deeper tubes under Washington and Jackson Streets for use by streetcars. Except near the river where these new subways would have to descend to greater depth than the freight tunnels to cross the river,

Chicago Historical Society (ICHi-23264)

Construction of a trunk tunnel segment on Franklin near Washington in 1902.

the subways were to actually rest on top of the tunnels or pass just above them.

While implementation of this plan would have resulted in the loss of some tunnels near the river, the majority of the system would have remained undisturbed. The city's planners evidently viewed the freight tunnel system as a utility that was to remain as free from dislocation as possible. By the 1930s, however, subway proposals simply called for the removal of tunnels when they were in the path of construction with no provisions to relocate or otherwise preserve them. This meant that instead of relocating a tunnel to allow for continued access to serve a customer, that business would simply be lost forever. Clearly, the tunnel system was no longer thought of as an important utility to be maintained at whatever the cost.

Finally, with reference to passenger service, the amended franchise ordinance declared that "...nothing herein contained shall authorize said Illinois Telephone and Telegraph Company to maintain or operate cars or vehicles of any kind in its said tunnels for the conveyance or transportation of passengers..." This, however, did not prevent the successor Illinois Tunnel Company from purchasing several such cars which were used for transporting special groups on tours of the system.

SCIENTIFIC AMERICAN

[Entered at the Post Office of New York, N. Y., as Second Class Matter. Copyright, 1910, by Munn & Co., Inc.]

A POPULAR ILLUSTRATED WEEKLY OF THE WORLD'S PROGRESS

Vol. CII.—No. 22.] ESTABLISHED 1845.
NEW YORK, MAY 28, 1910.
[10 CENTS A COPY. $3.00 A YEAR.

First section of subways to be undertaken.

Chicago proposes gradually to place all tracks, surface and elevated, that enter the business zone, below street level. The above view shows a station on the new Wabash Avenue four-track subway. The trolley cars are to the right, the elevated express cars use the left-hand track. The lowest tunnel is the present 60-mile freight subway.

HOW CHICAGO IS SOLVING ITS RAPID TRANSIT PROBLEM.—[See page 439.]

Bruce G. Moffat Collection

Larry Best Collection

General Manager and Chief Engineer George W. Jackson hard at work in 1905. On his desk can be seen two of the IT&T's Strowger-type dial telephones.

George W. Jackson

The man in charge of overseeing the building process was a remarkable individual named George W. Jackson. Born in Chicago on July 21, 1861, he was an Oxford-educated civil engineer specializing in tunnels, bridges, sewers and conduit systems. Jackson also served as a consulting engineer for the City Council's Committee on Local Transportation which was considering the construction of downtown streetcar subways and the potential for municipalization of the city's several street railway companies.

In addition to being an officer and investor in the Illinois Telephone & Telegraph Company, Jackson owned and operated several businesses including a successful engineering company and two steel fabrication concerns.

He also had a remarkable flair for self-promotion, making sure that his name and accomplishments were put before the public at every opportunity. In 1910, he published a book called "Subways" which promoted his plan for a downtown streetcar subway system. The book included a copy of his 1892 charter for the Chicago Underground Subway Company. Although his downtown subway plan, like countless others, failed to materialize, the book contained numerous examples of engineering and construction projects that he and his engineering firm, George W. Jackson, Inc., had undertaken. Among the projects illustrated were several water tunnels and sewers, railroad viaducts, schools and bridges.

But undoubtedly his largest single project was the construction of virtually the entire freight

Left: *In 1910, Scientific American published this transportation plan developed by City Engineer John Erickson calling for a streetcar subway to supplement the city's famous elevated system. At that time, preservation of the freight tunnel system, was a given.*

Jackson spent considerable time below ground supervising the work. In this circa-1900 view, he is seen standing at right in what is probably the LaSalle Street tunnel at its intersection with the alley bore leading to Shaft #1. Note the Automatic Electric-supplied construction telephone on the wall.

Bruce G. Moffat Collection

tunnel system. Although it is unclear whether he was involved from the time that the first shaft was sunk in the basement of the saloon, he was actively managing the project by 1901. According to a 1904 news account, the building of this extensive network of tunnels required a labor force of 1,200 men divided into three shifts working at sites scattered around the downtown area. The engineering and design of the bores and the on-going inspection of the construction work and direction of the laborers was an especially daunting challenge.

On August 7, 1904, at the height of tunnel construction, the *Tribune* published a particularly interesting, and lengthy, "news" article trumpeting his tireless devotion to the tunnel project and the novel method he used to juggle work and family obligations. Among the more interesting passages were these:

The Busiest Man in This Hustling City

George W. Jackson, the man who is digging the tunnels that will make Chicago a city with underground transportation in the downtown districts, is so busy that for the last three years he has found little time to spend with his family. It is necessary for him to devote his time night and day to the management of the affairs of which he is the superintendent, and when he did occasionally steal a few hours from his work to become acquainted with his family it was more than probable a jangling telephone would tear him away from his hearthside and drag him out to do something in a downtown tunnel, leaving his wife and family wishing that tunnels or anything like them had never been invented.

This statement of affairs continued until Mr. Jackson decided that it was to be borne no longer. He could not leave his work or turn it over to the supervision of some one else, and he could not take it with him to his home. So he resolved to do the next best thing, and take his home to his work. When the Illinois Tunnel company's offices were moved into the new building in Monroe street [from leased space on LaSalle Street to 177 Monroe street; a short distance from the company's main telephone switching facility on Fifth Avenue] Mr. Jackson had the top floor fitted up as an elegant modern flat, and now has his family where he can be with it without leaving his work.

An obviously retouched photo showing George W. Jackson (at left) supervising construction of a large diameter bore beneath the city's main post office in 1905.

Larry Best Collection

Works 24 Hours a Day

If there is another man in the city who works as much or as hard as does Mr. Jackson, his associates would like to meet him. While there are many men who spend twelve or fourteen hours of the twenty-four at their places of business and work hard all the time they are there, Mr. Jackson is at the beck and call of duty every hour in the twenty-four and often works the entire time at a stretch. All the while that he has been superintending the building of the downtown tunnels he has been thus occupied. Night and day he is with his work; it is at his table and his bedside. Most of his time is spent underground, so night or day makes little difference to him...

The time that he managed to spend with his family was only such as he was able to snatch between calls. Sometimes it was days when he did not see either his wife or his children, and then often when he left the office and went to his home it was only to find a telephone summons awaiting him with the information that his presence was needed downtown. With only a greeting to his family he turned back to work. There was no disobeying the summons that came from underground...

Jackson's new apartment had 10 rooms and was lavishly decorated. To ensure that he was never out of contact with his subordinates, he had six telephones scattered throughout the apartment. (Presumably he didn't need all of those telephones after the cessation of most construction in 1910 and his retirement from active involvement in the company's affairs.)

Even with all of his pressing business commitments, Jackson managed to find time to belong to various civic and fraternal organizations including the Western Society of Engineers, Illinois Athletic Club, Knights Templar, Elks and Masons. He passed away on February 5, 1922.

SECTION
II
The Illinois Tunnel Company

3 Getting Down to Business

Once the amended franchise was approved construction of the railroad and telephone systems was accelerated to meet the new deadline. As construction activity picked up, the financial needs of the enterprise likewise increased, making it necessary to obtain additional working capital. This was accomplished by incorporating a new company with broader corporate powers and a larger capitalization to take the place of the IT&T.

On October 29, 1903, articles of incorporation for the new Illinois Tunnel Company were filed. The company's organizers were listed as Charles C. Wheeler (brother of Albert G.), IT&T attorney Henry A. Wilkening, and Thomas A. Moran Jr., whose father was the new corporation's attorney. The promoters attempted to keep the identities of their investors as secret as possible. But on November 11, the *Journal* reported that an Edwin Gearhart had purchased stock valued at $2,999,300. This represented all but 7 of the company's 300,000 shares. Speculation was rampant as to whom Gearhart was representing. An answer of sorts to that question did not come until late January of 1904, when the *Tribune* reported that the company was being financed by J.B. Russell & Co. of Wilkes-Barre, Pennsylvania, for an unnamed eastern anthracite coal syndicate. It was also stated that Russell had earlier purchased all of the IT&T's bonds and for a while had served as the company's secretary.

In its application to the State of Illinois for incorporation, the Illinois Tunnel Company stated its objectives in the broadest possible manner to maximize the company's profit making potential and to improve the marketability of its securities. Specifically, the company declared its intention to "furnish, transmit, convey and deliver signals, sounds, intelligence, packages, mail matter and general merchandise, power, heat and light by steam, water, air, electricity and otherwise, and to acquire, construct, dispose of, hold, maintain, operate and lease to or rent from others all tunnels, instruments and appliances and other property, real or personal, in carrying out such objects." Although the transmission of heat and power through the tunnels was not included among the activities allowed under the franchise inherited from the IT&T, it was clear that the company's leaders felt that such uses might be feasible. Interestingly, although operation of a telephone system was implied, it was not explicitly stated, reflecting the ongoing shift in corporate interest in favor of railroad operations.

Left: *In 1902 members of the Illinois Telephone &Telegraph's engineering and surveying crews posed for the camera in one of two short trunk tunnel segments that had been built under Franklin between Washington and Madison.*

Larry Best Collection

Following incorporation, the new company issued $30 million in mortgage bonds. $5 million was used to acquire the properties and franchises of the IT&T (including control of the Illinois Telephone Construction Company) and to retire the original bonds.

The remainder was to be used to expand the system from the existing 20 miles of tunnel to the 50 miles mandated by the city. Wheeler went one step further and proclaimed that he would build 60 miles. His plan included a link to the manufacturing and food-processing plants situated at the Chicago Union Stock Yards about five miles southwest of the Loop. This particular extension was never built, although the company did manage to build and operate about 60 miles of tunnel in and around the Loop area.

The Illinois Tunnel Company officially took over from the IT&T on January 28, 1903, when a deed transferring the old company's property was filed with the county recorder. The IT&T continued to exist on paper, however, until December 12, 1904, at which time its dissolution was approved by the stockholders. The vote was strictly a formality as the company's 50,000 shares were represented by the company's four directors: President Albert G. Wheeler and J. B. Russell (49,800 shares combined), Albert G. Wheeler, Jr. (100 shares), and Joseph Harris of the Automatic Electric Company (100 shares).

The First Locomotives Arrive

In late 1903, three 24-inch gauge mine-type electric locomotives were received for testing to determine what type of haulage system should be used in the completed tunnels for the handling of freight. Two were supplied by the Morgan Electric Machine Company of East Chicago, Indiana. The Morgan locomotives collected current from an energized center cog rail which was used to gain traction. The third locomotive was a conventional adhesion-type mine locomotive built by the Jeffrey Manufacturing Company of Columbus, Ohio. This locomotive received traction power through a trolley pole that made contact with an energized overhead wire.

The exclusive use of the Morgan system was seriously considered for a time before being dropped in favor of conventional adhesion-type mine locomotives. Evidently, the grades at the river crossings did not pose as great an operating challenge for adhesion locomotives as had been expected. As a result, the use of mine locomotives collecting current from an overhead trolley wire won out over the Morgan cog rail system which, among other things, required relatively elaborate switches if it were desired to maintain an uninterrupted cog rail at turnouts.

The Morgan system was, however, used on a limited basis to serve the company's newest disposal station located along the shore of Lake Michigan in what is now Grant Park. The cog rail technology was needed here to enable the locomotives to haul the heavy loads of spoil to the surface for dumping. Much of today's Grant Park was claimed from Lake Michigan using spoil from the construction of the freight tunnels. All other train operations utilized trolley locomotives supplied by a variety of manufacturers, including General Electric, Baldwin-Westinghouse, Goodman and Jeffrey. The average locomotive weighed 5 to 7 tons and was limited to handling trains of up to 15 cars.

The March 15, 1904, issue of *Electrical Mining* carried a lengthy article about the installation of the Morgan technology in the tunnels. The article stated in part:

> The Goodman Manufacturing Co., Chicago, sales agent for the products of the Morgan Electric Machine Co., of East Chicago, Ind., has supplied two of the Morgan third-rail locomotives and has equipped two miles of experimental track with the necessary third-rail appliances and connections for full test of the Morgan System under the actual conditions of the required service. In this system of haulage a combination rack and conductor rail is placed between the track rails and fulfills the double purpose of conveying the electric power current and of serving as a rack with which the driving sprockets of the Morgan locomotive enmesh.
>
> The track rails weigh 56 pounds per yard and are laid to a gauge of 24 inches on wooden ties resting upon the concrete floor of the tunnel. The third rail is made up of heavy flat bars of iron or steel, perforated with square holes to admit the locomotive sprocket teeth... The third-rail in the tunnels is carried by insulating wooden stringer pieces and encased by strips of wood which leave the rail uncovered only...sufficiently to admit the teeth of the driving sprocket.

Goodman Equipment Corporation

In 1904, the Goodman Manufacturing Company, a manufacturer and distributior of mining equipment began publishing its own magazine Electrical Mining. The cover of the first issue featured an Illinois Tunnel train. Although the locomotive was not built by Goodman, they did sell Morgan products.

About the Morgan locomotives, *Electrical Mining* had this to say:

> Two of these machines are now in use. They are practically duplicates, each weighing 6,000 pounds and rated at 80 horsepower. The haulage speed in regular service is designed to be about ten miles per hour, at which rate each locomotive is expected to draw 12 or 15 loaded cars. The current is received from the third rail by the sprocket teeth and passes thence through spring contact brushes, bearing against the side of the sprocket to the conductors leading to the motor armature.

Morgan would supply at least two additional locomotives once landfill operations got underway.

Zenon Hansen Collection

An early 1905 view showing two different types of Morgan locomotives hauling test loads of freight prior to the start of revenue service. Note the differences in headlights and mounting of controllers.

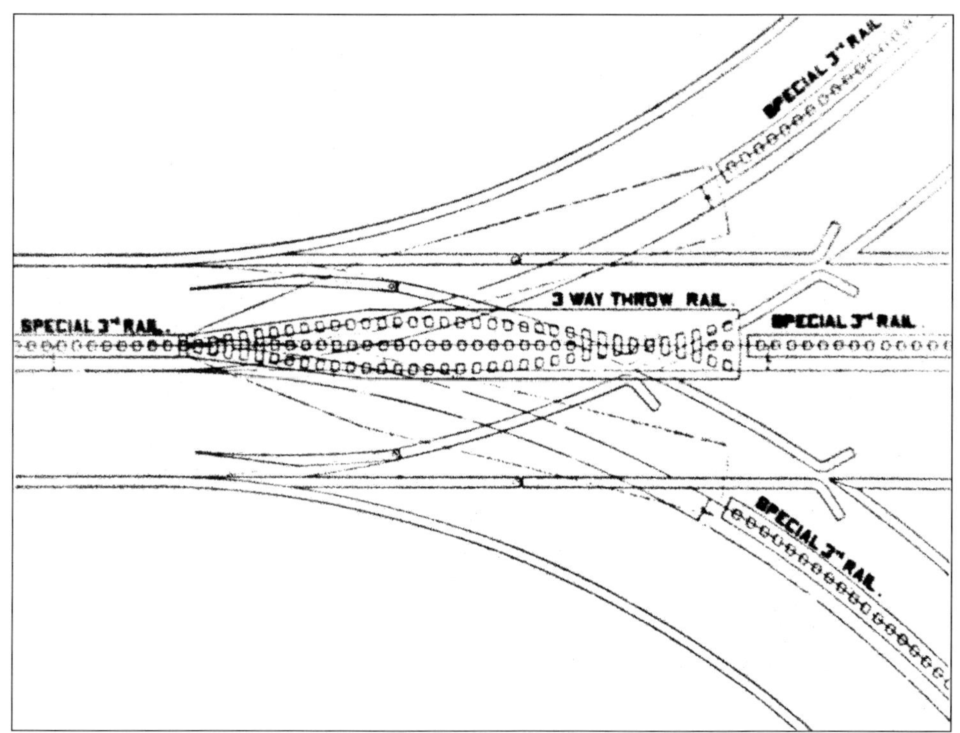

Goodman Equipment Corporation

Detail of a typical three-way third rail switch.

Building Chicago's Front Yard

When construction of the tunnel system began, most of the great expanse of downtown park land that today extends east from Michigan Avenue did not exist. In fact, dry land generally ended at the Illinois Central tracks (in some places, the tracks were actually built on a trestle built into the lake). The city's South Park Commissioners had a plan to create a park by creating a giant landfill. The building of what would eventually be referred to as "Chicago's front yard" was being championed by Chicago mail order tycoon A. Montgomery Ward, who tirelessly fought all attempts to construct public buildings in the projected park area east of Michigan Avenue and south of Randolph street. In a typical example of bureaucratic creativity the as-yet largely non-existent park was initially referred to as Lake Front Park but was soon formally renamed Grant Park. At this time the two names were used interchangeably.

Now all that the park commissioners needed was spoil to dump in the water. Fortunately, their desire coincided with the Illinois Telephone Construction Company's need to find a new site for disposing of the seemingly limitless quantities of dirt and clay being removed by its tunneling crews and from building construction sites where the tunnel trains would be used to handle the spoil removal duties. On May 11, 1904, the commissioners accepted the company's offer of providing free spoil and awarded them a contract to supply more 1.24 million cubic yards of fill at no charge by April 1, 1907 (this figure was soon increased to 4.1 million yards). At the same time, the commissioners also awarded a contract to the Chicago & Great Lakes Dredge & Dock Company to supply an additional two million yards of spoil gained from commercial dredging operations in area waterways, at a nominal price of 10 to 17.9 cents per yard.

Besides providing free spoil, the company's public spirited promoters also had to bear the entire cost of building a ramp to bring the trains to the surface as well as all equipment and manpower necessary to

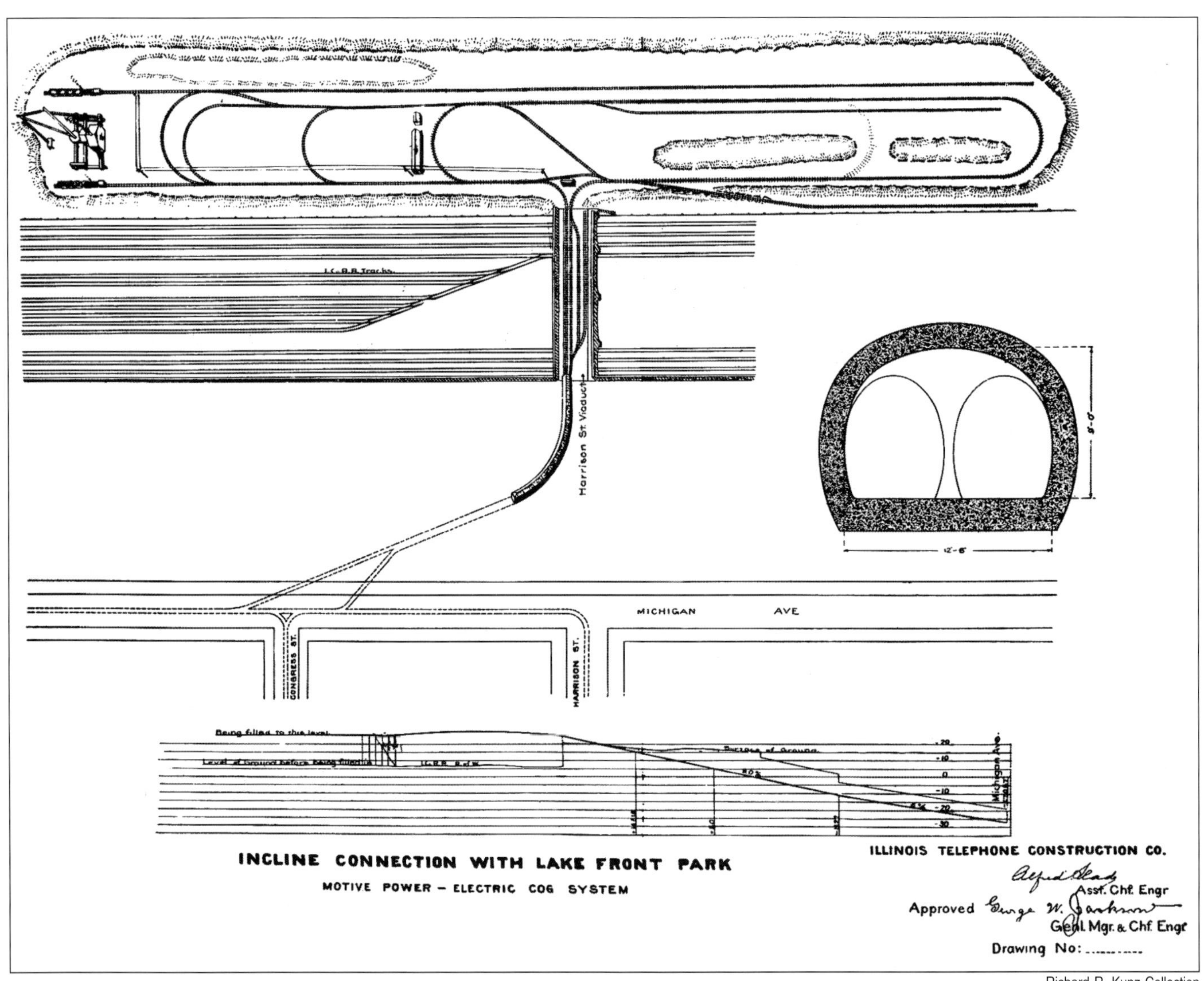

Richard R. Kunz Collection

Plan showing the layout of the Grant Park disposal station trackage in 1904.

Bruce G. Moffat Collection

Looking southeast near Congress and Michigan. Most of the ramp connecting the subterranean trackage with the large shed in the background was built by excavating down from the surface.

Larry Best Collection

Another view of the ramp work looking northwest. The Art Institute is in the right background.

Chicago Historical Society (DN-5233)

The shed in the foreground served as the transition point between the subterranean and surface trackage at Grant Park. Originally the tracks on the bridge spanning the Illinois Central Railroad unprotected from the elements. However by the time of this 1907 photograph a roof had been added. Landfill operations were nearly complete.

distribute the spoil and compact it. Although this work would require a considerable expenditure with no revenue being generated, Wheeler and his associates apparently believed that the resultant publicity would be beneficial and prove to a skeptical city government that the company was not trying to cheat them in some way. (The company's activities were frequently being assailed by opportunity-seeking aldermen and the local "reform" establishment who felt that Wheeler and his out-of-town backers were cheating the city out of money rightfully due it for allowing the company to utilize street rights-of-way for commercial gain.)

On June 2, 1904, a pile driver started work on installing the sheet pilings (a new construction innovation for which Jackson claimed partial credit) for the two track ramp which was to have a gradient of 9%. This extremely steep gradient was needed so that the tracks could rise approximately 40 feet to the surface in as short a distance as possible. For this reason, the company settled on using the Morgan cog rail locomotives on the spoil trains even after it was determined that conventional trolley locomotives would be suitable for regular freight service.

The Morgan system had already been installed in a number of tunnels including those under Fifth Avenue, Washington Street and Michigan Avenue. To get the trains into the park the Illinois Telephone Construction Company's forces built a pair of converging inclines that left the Michigan tunnel on either side of Congress. Near the surface the two tracks entered a curving wooden shed leading to a bridge (eventually enclosed) spanning the Illinois Central Railroad tracks. Temporary surface tracks would be used to bring the spoil cars to the desired locations where they would be unloaded using steam-powered derricks.

Barely had work on the ramp begun then Ward lodged a lawsuit to stop the construction. Although he was not against the construction of the landfill, he did object to the company's plan to build the ramp and shed which he maintained would be both an eyesore and a violation of an 1896 injunction forbidding the construction of buildings in the park. The judge disagreed and allowed the shed and ramp to be built with the stipulation that they had to be removed once the landfill work was finished.

Finally, on October 5, 1904, the first spoil train

Chicago Historical Society (ICHi-23268)

A Morgan locomotive demonstrates the advantages of cog rail technology as it pushes a string of Newman-type dump cars up the steep ramp leading to the Grant Park landfill in late 1905.

climbed the ramp into Grant Park. South Park and Tunnel Company officials were on-hand to celebrate this event as were a contingent of alderman and other interested citizens. Ten days later the first loads of spoil removed from a commercial building construction site arrived by train. This material came from the site of the Heyworth Building (located at the southwest corner of Madison and Wabash) and required the use of two dedicated 12-car trains that shuttled back and forth. This was apparently the first commodity handled by the tunnel trains on a for-hire basis. The building of Chicago's "front yard" was now officially underway.

Transportation of construction spoil was now largely accomplished using two-foot gauge cars of two basic designs. The most common of these was referred to as a "box" car. This design utilized a flatcar-type frame, on which was placed an open-topped wooden box having a hinged floor panel. The other car design was the so-called Newman-patented dump type, which vaguely resembled a tilt-body gondola. Many of the box cars were reassigned to ash haulage as the spoil traffic declined.

Once the cars were filled animal power (or in some cases, trolley locomotives) was used to move them to a point along the cog rail trackage where one of the Morgans took over. In at least a few cases the older 14" gauge construction cars were used but this required them to be loaded on two foot gauge flat cars for the trip to the park. Once at the bridge spanning the Illinois Central tracks, the entire train continued into the landfill site on a trestle where their loads were simply dumped onto the ground. This practice was soon revised so that the electric locomotives were uncoupled on the bridge and replaced by one of the Illinois Telephone Construction Company's two Porter-built 0-4-0 saddle tank steam locomotives. The steam locomotives then towed the cars onto temporary trackage laid on the ever-growing landfill and spotted them under one of two stiff-legged steam derricks. The derrick would lift the box off the car frame and position the car. By means of a lever attached to a long chain or rope, the bottom of the car was then opened, allowing the contents to be deposited in the desired location (the Newman cars were not handled in this manner since they had opening side panels).

Western Electrician

A Morgan poses with a flat car carrying two loaded 14" gauge construction cars on the bridge spanning the Illinois Central tracks. This view looks northwest towards Michigan and Congress.

Chicago Historical Society Collection (ICHi-23275)

For a brief period the third rail trains ventured well into the new park area, which the company unceremoniously referred to as Shaft #11 or "Lake Front Dump." The work crew accompanying the two motor Morgan locomotive has emptied the two Newman dump cars and 10 of the 14" gauge construction cars (carried aboard five flat cars) of their spoil.

Wheeler Sees A Media Opportunity

Ever the promoter, Wheeler saw an opportunity to garner some positive publicity for his scandal-plagued company by inviting reporters to observe and photograph the first train loads of spoil being dumped into Lake Michigan. In this way the general public would know that the Illinois Tunnel Company was concerned about its host city and was playing a major role in building what would become its greatest park. Typical of the articles written about that October 5, 1904, "ground dumping" is this one which appeared in the *Inter Ocean* the following morning:

Larry Best Collection

On October 5, 1904, a Chicago Inter Ocean *photographer recorded Henry G. Foreman (at left) of the South Park Commission joining Illinois Tunnel Company officials Albert G. Wheeler, George W. Jackson and S. A. Kitchener to view the beginning of the landfill work from one of the company's inspection cars.*

GRANT PARK IS STARTED

BEGINNING OF FILLING IN WORK BEGUN WITH CEREMONY

President Albert G. Wheeler of the Illinois Tunnel company assisted by President H. G. Foreman of the South park commission, seized a heavy sledge hammer and with two or three well directed blows knocked loose the iron bolt holding closed the door of a car filled with dirt yesterday, sending its contents splashing into Lake Michigan. And as the spray and foam from the several tons of earth mixed with the water, a great shout went up from the several hundred people present at the foot of Harrison street, with the South park commissioners and officials of the tunnel company, local railway companies, and members of the city council.

It was the beginning of the end of the completion of Grant park. One hundred cars of dirt were dumped in during the afternoon, and it will require 4,000,000 cubic yards of earth before the resort is ready for use. It will require about two years to complete the work.

When finished the park will be one and one-quarter miles long and an eighth of a mile wide – east of the Illinois Central railroad tracks.

And it will be made the finest city resort contiguous to business districts possessed by any city in the world.

It will contain besides the magnificent Field museum, the Art institute and the new Crerar library. Plans are made for widening Michigan boulevard for linking the new Lincoln park and South park systems.

This filling in is now done without cost to the taxpayers. This means a savings of half a million dollars, the lowest bid for the work required.

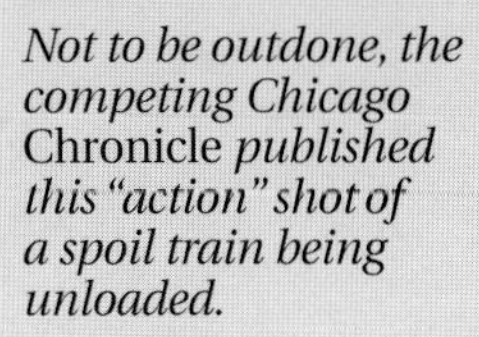

Not to be outdone, the competing Chicago Chronicle *published this "action" shot of a spoil train being unloaded.*

Larry Best Collection

The Heyworth Building

Bruce G. Moffat Collection

On November 22, 1904, an inspection party visited the subbasement of Chicago's newest skyscraper - the Heyworth Building. Located on the southwest corner of Madison and Wabash, this office building had the distinction of being the first Loop building project that utilized tunnel trains to remove construction spoil. Note the contrasting exterior appearances of inspection cars 30 and 31.

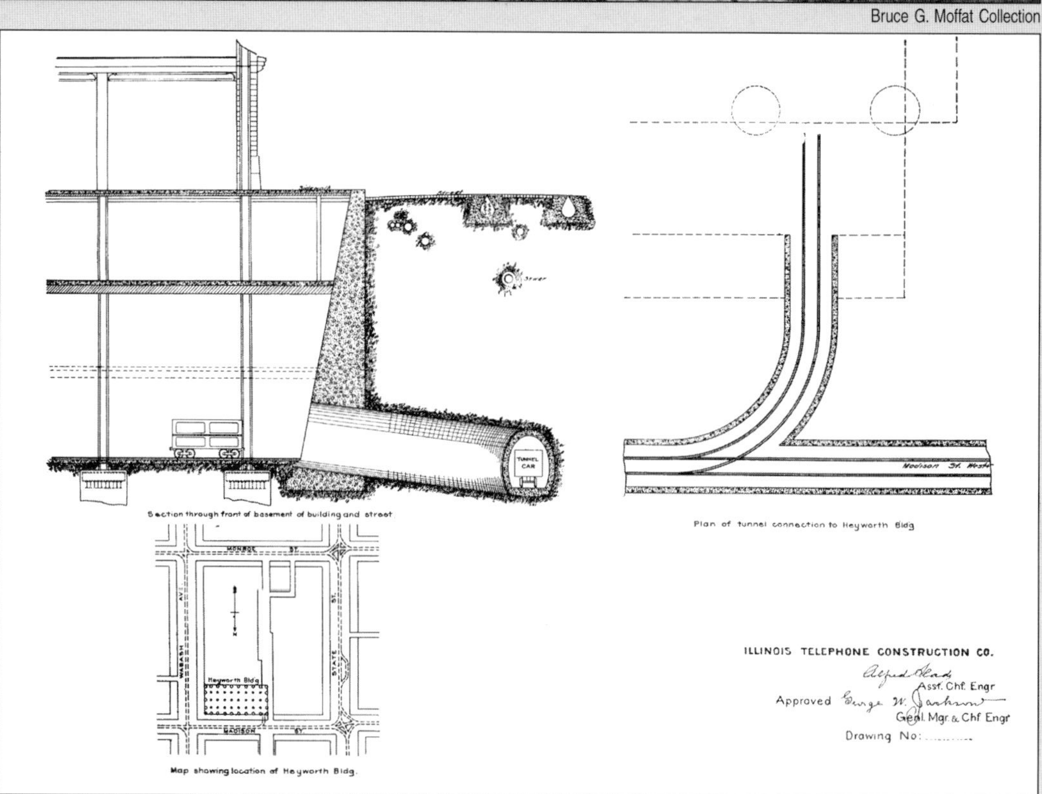

Site plan showing the building's connecting spur from the Madison tunnel.

J.J. Sedelmaier Collection

Western Electrician

A motorman prepares to couple his General Electric-built locomotive onto a loaded Newman-type dump car. The car will be then be parked in a nearby tunnel for pickup by one of the Morgan locomotives for delivery to the Grant Park dump site.

Larry Best Collection

The building's engine room as it looked in its final form.

Larry Best Collection

Note the ash car parked behind the coal pile.

Chicago Historical Society Collection (ICHi-31237)

By the time this photograph was taken on October 23, 1905, the Morgans no longer ventured east of the Illinois Central bridge. Beyond the bridge, the loaded cars were towed by steam locomotives to the dump site. In this view looking east, one of the steam locomotives is ready to enter the landfill while the operator of 110 (center) gets ready to return underground. Meanwhile, the operator of 109 makes some adjustments to his locomotive.

Locomotive 19 stands by while a derrick lifts a loaded box of spoil and positions it for dumping. The date is October 22, 1905.

Bruce G. Moffat Collection

Another view showing both of the Porter locomotives in action.

Bruce G. Moffat Collection

Chicago Historical Society Collection (ICHi:31232)

During 1904 and early 1905 simulated freight operations were conducted to gain experience. On January 2, 1904, the company photographer recorded one of these trains at an unknown location.

Non-Revenue Operations are Begun

With about 20 miles of the two-foot gauge system completed, train movements on a limited scale were initiated on January 2, 1904. The Chicago Edison Company furnished the 250-volt D.C. traction power used by the locomotives. Ultimately the system drew its power from four different Edison substations. Use of purchased power was envisioned to be only temporary as the tunnel company planned to build its own power station on leased property along the Chicago River at 24th Street. Purchased power proved to be less expensive however, resulting in the lease on the power station site being allowed to lapse.

The first movements were handled by one of the two Morgan locomotives. The first train consisted of several general merchandise cars that had arrived on the property during the preceding year for testing. The cars were loaded with barrels to simulate anticipated operating conditions. Although this event received no attention from the general press, the company photographer was present and recorded several views that were later reproduced in industry trade journals. Two views were even published as post cards and placed in general circulation.

At this time normal car loadings consisted mainly of clay and dirt spoil mined at the various tunnel construction sites. These trains were technically under the control of the Illinois Telephone Construction Company which was in charge of the construction work even though the locomotives and cars were lettered for the parent Illinois Tunnel Company.

In late 1904 and 1905, test shipments of coal and general merchandise freight were handled. This was done apparently to determine if the Morgan system was more suitable for general haulage than standard adhesion-type mine locomotives. These tests also allowed management to gain needed experience in running the railroad. The extent to which the Morgans were used in this capacity is not known, but it was probably very limited as few building connections had been completed. Formal revenue freight operations would not begin until 1906.

The Morgan locomotives remained in service through at least 1907, ending only when the Grant Park inclines were closed following the cessation of landfill operations north of 12th Street. The cog rail was then removed and all tunnel trackage was worked by trolley-equipped locomotives. The Morgans were apparently scrapped.

177-179 Monroe Street — Below Ground

The Tunnel Company's office building at 177-179 Monroe (between LaSalle and Wells) housed a variety of activities besides the usual administration and engineering activities. George W. Jackson had the top floor converted into a luxurious apartment suite for him and his family while the basement housed a temporary maintenance facility for the growing fleet of locomotives.

A loaded spoil car is being lowered into the tunnel from basement level while guests of the company look on from one of the inspection cars that has been positioned on the turntable.

Larry Best Collection

A 1905 view of one of the Jeffrey's being tested in the basement.

Zenon Hansen Collection

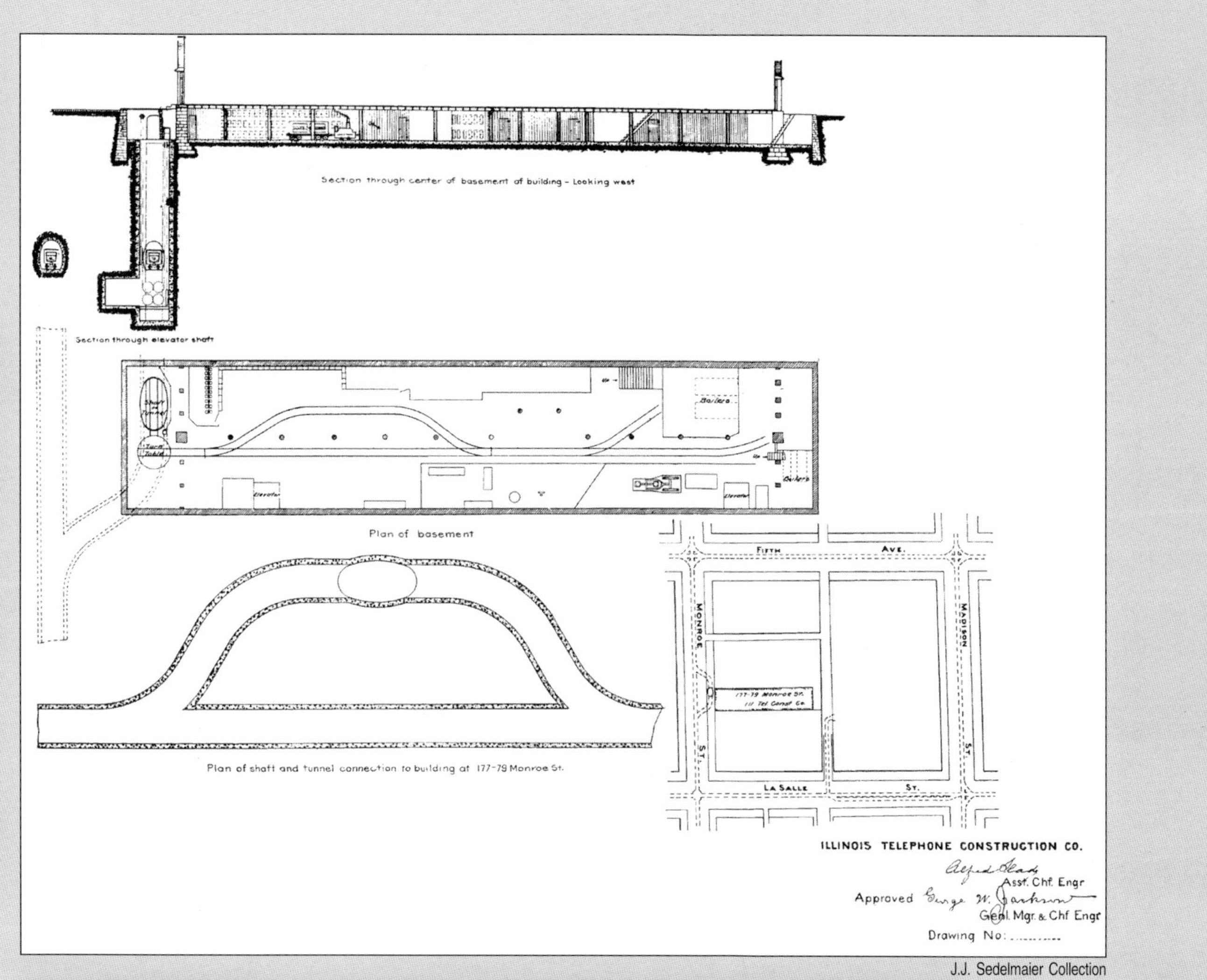

J.J. Sedelmaier Collection

A plan showing the track layout. Note the turntable and elevator that were needed to connect this track with the tunnel below.

Chicago Historical Society Collection (DN-003084)

On March 13, 1906, a delegation of city officials visited the facility as part of an official inspection tour of the nearly completed system. General Manager George W. Jackson (at right) looks on as Mayor Edward F. Dunne (standing next to locomotive) turns to make a comment to city electrician William Carroll. Although perhaps only a year or so old, the Jeffrey locomotive appears to have seen heavy use. The partially visible fleet number is a bit of a mystery since it does not correspond with surviving data.

177-179 Monroe Street — Above Ground

Assistant treasurer J. C. Law had his office on the ground floor.

Larry Best Collection

Several floors above street level was the company's drafting department.

Larry Best Collection

Larry Best Collection

The chief draftsman's office was marked only by a railing which enabled him to readily monitor his employees.

Chicago Historical Society Collection (ICHi-23284)

The furnishings found in George W. Jackson's office reflected his stature within the organization.

Larry Best Collection

Jackson's outer office staff occupied these well-appointed quarters. Note the sky light.

The banquet in the Jackson tunnel on February 10, 1904, was a unique dining experience to say the least.

Bruce G. Moffat Collection

In Search of Customers

From all indications, the Illinois Tunnel Company's promoters concentrated on building the tunnels first and looking for customers later. This was at best a risky approach that resulted in miles of tunnels being built without any building connections at all. It was not until the tunnels were largely completed in a given area that contracts for installing connections to the adjacent buildings were actively solicited.

Company officials apparently believed that having as much of the system built and in operation as possible increased their chances of attracting customers. They may have also wished to avoid possible opposition from the teamsters and adjacent property owners. The teamsters and their powerful union would likely object to the loss of any business (and jobs) where they had previously enjoyed a monopoly. On the other hand, building owners were more concerned about potential structural damage caused by the tunneling activity.

An example of the latter concern is illustrated by this account involving Marshall Field who operated the city's premier department store. In its May 14, 1906 issue, the *Railroad Gazette* reported that Field, upon learning that construction was being undertaken near his store, vowed to take legal action to halt the work. He was too late, however. The work having been completed for quite some time:

> Mr. Field was invited down to visit the streets of a city he had but recently learned about. As he read the brass signs on the street corners below he exclaimed: "So this is my first view of my own corner underground. Can I see the other three corners?" In a few minutes he had been upon a tour of inspection about the other three corners and around his own block. Mr. Field stood and looked down the long alley, through whitewashed and electrically lighted walls relieved in color only by the lead pipes filling the arched roofs and carrying hundreds of miles of telephone cables connecting the 8,000 or 10,000 subscribers to the Chicago Automatic Telephone system. He reflected for a few minutes and then exclaimed: "Well, it was fortunate I did not know what you were doing down here. For I certainly would have fought you in the courts. Real estate may be as valuable under ground as above ground. You need not bother to build elevators under our sidewalk. I will meet you more than half way. I will come down to your level with the entire building."

This was soon accomplished, with several floor-level entrances being built into the building's second sub-basement. In addition to track, overhead wire was also installed, permitting the locomotives to spot cars within the building.

Dinner is Served

To commemorate the start of train operations, a banquet was held for members of the Chicago Press Club and their guests on February 10, 1904. Although banquets were not an uncommon way to mark the openings of new rail lines, this event was to be unlike any other. In fact, even the banquet site was unique. Instead of renting a banquet hall or using its office facilities, the Illinois Tunnel Company hosted the event in the Jackson Boulevard tunnel.

On the night of the big event, the dinner party, numbering more than 600 by one estimate, entered the system through the company's automatic telephone switching facility at 177 Fifth Avenue (now 105 S. Wells Street) and descended to tunnel level by elevator. There they were met by a special train which transported them on a short tour of part of the system and deposited them at the banquet site, centered at the intersection of Jackson and Fifth.

To convert the tunnel into a banquet hall suit-

Larry Best Collection

Members of the city's various women's clubs visited the tunnels on April 15, 1904. Note the rectangular brass "builders marker" on the wall at left. Less obvious is one of the brass street signs located to the right of the group.

Larry Best Collection

Close-up view of the marker. Installed at various locations, these large brass signs were removed after only a few years.

Ed Anderson

A typical brass street sign of the period. These were later replaced with stenciled legends.

able for this momentous occasion, the ingenious management installed temporary wooden flooring and a line of tables extending two city blocks from LaSalle to Franklin. Temporary overhead lighting and decorative bunting were used to offset the white concrete walls and provide a more festive appearance. In addition, some of the company's new dial telephones were placed on the tables at 30-foot intervals to give Press Club members the opportunity to test the latest in modern communications equipment. As a final touch, an orchestra was stationed at the LaSalle Street end of the tables to play background music. It must have been quite an experience to listen to the musical selections as they reverberated off of the concrete walls. Oddly enough, none of the numerous accounts of this event made mention of what was on the menu.

A *Chronicle* reporter did note, however:

> The novelty of the affair appealed strongly to the small army of guests that tramped east and west in the tunnel to find seats at the table. It proved anything but a welcome novelty, however, to the 300 waiters engaged for the occasion, for they had to walk half a block between the serving of each course in the dinner.

Following dinner, Tunnel Company President Albert G. Wheeler and General Manager/Chief Engineer George W. Jackson spoke about the building of the system and their expectations for its success. It was revealed that about $7 million had been spent on its construction up to that time and that that the railway's fleet consisted of three locomotives and 24 freight cars. In a fit of corporate exuberance, it was announced that there would be 150 locomotives and 2,000 cars in service by June 1. Wheeler also remarked: "London, Paris, and other cities of the old world transport their passengers through tunnels under the streets. Chicago reverses this order of things and the people are to be kept on the streets, where they can enjoy the fresh air, while the freight traffic is to be sent through these tunnels." Unfortunately the lack of acoustics and the two-block length of the banquet "hall" prevented all but those closest to the speechmakers from hearing what was said.

For those who wished to inspect more of the system than had been seen by train on the way to the banquet site, uniformed guides were available to give walking tours. The more adventurous members of the Press Club were permitted to explore as much of the system as they wished.

A Popular Tourist Destination

A visit to the world's only urban tunnel freight railway was high on the list of activities that persons having professional credentials, business connections, or some other novel justification, sought to accomplish while visiting the city. Of course the tunnels were not opened to the general public, and perhaps because of the company's somewhat rocky relationship with the city, no attempt was made to get permission for the operation of public tour trains. Nevertheless, Jackson found time to escort delegations of government officials, captains of industry, financiers, and engineering delegations through the underground maze. Many of these visits were publicized by the company for public relations purposes.

Perhaps one of the most novel tours was given on April 13, 1904, when Mr. and Mrs. Thomas C. Ashecroft, a honeymooning couple from Memphis, were treated to a ride. As with most other tours, the couple reported to the company's office and telephone exchange at 177 Fifth Avenue where they boarded their special car which was then lowered by elevator into the tunnel and coupled onto one of the company's two third-rail locomotives for a trip down Monroe Street. Jackson was quoted in the *Inter-Ocean* as saying: "The young couple enjoyed the trip immensely, the tunnels being brightly lighted with incandescent light. I want all the brides in Chicago to try it." There is no record if any did.

Two days later, a delegation of 17 women representing the city's premier women's clubs and neighborhood organizations were Jackson's guests for a two hour tour. The *Post* reported: "The tunnels, lighted with electricity and the absence of smoke, dust and dirt, attracted the attention of the women and was responsible for much favorable comment."

One group that was not welcome was the Chicago Federation of Labor. The labor organization's leaders were upset that their members stood to lose business to the trains. They were also unhappy that most of the workmen toiling at the construction headings beneath the streets were not union members and they harbored hopes of organizing them. Whenever the labor organizers made an inquiry to inspect Chicago's subterranean engineering wonder, Jackson and the other company officials were always "too busy" or said that the narrow confines of the tunnel and the ongoing construction activity made a tour ill-advised from a safety standpoint.

While the Chicago Federation of Labor never got its tour, the system's train operators and warehousemen were eventually organized by the Brotherhood of Railway & Steamship Clerks.

Larry Best Collection

Another view of the "women's club" tour. Note that this track segment is equipped for both third rail and overhead trolley operation.

The Chicago Subway Company

On November 21, 1904, the first of several holding companies in tunnel history was incorporated in Trenton, New Jersey. The Chicago Subway Company, with a capitalization of $50 million, was established to arrange finances for, and own, the Illinois Tunnel Company and the Illinois Telephone Construction Company. From this time forward, the corporate structure of the affiliated companies would always include a holding company holding all stock, with individual investors being restricted to owning stock in the holding company. For all practical purposes, the subsidiaries were always operated as a single unit, sharing the same office space and personnel.

The Subway Company's board of directors read like a "Who's Who" of the business world. Among its members were such notables as: Edward H. Harriman, who controlled the Union Pacific, Illinois Central and Chicago & Alton railroads; James Stillman, president of the National City Bank of New York; V. A. Valentine, a director of Chicago meatpacking giant Armour & Co. and J. Ogden Armour's personal representative; and Albert G. Wheeler, president of the Illinois Tunnel Company. Following this, a number of railroad presidents were added to the Illinois Tunnel Company's

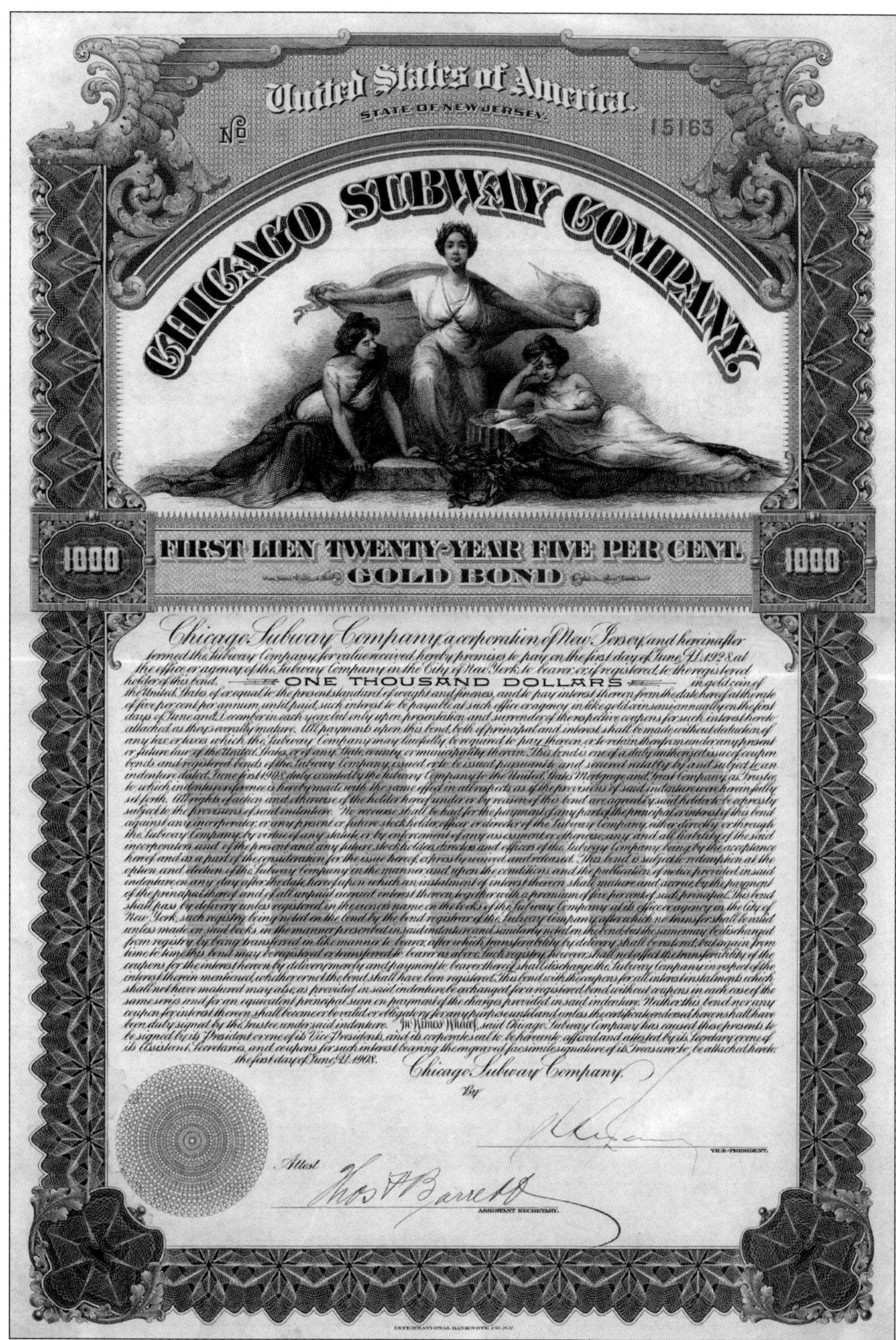

United States of America.
STATE OF NEW JERSEY.
No 15163

CHICAGO SUBWAY COMPANY

1000 FIRST LIEN TWENTY-YEAR FIVE PER CENT. GOLD BOND 1000

Chicago Subway Company, a corporation of New Jersey, and hereinafter termed the Subway Company, for value received, hereby promises to pay on the first day of June A.D. 1928, at the office or agency of the Subway Company in the City of New York, to bearer, or if registered, to the registered holder of this bond, ONE THOUSAND DOLLARS in gold coin of the United States of or equal to the present standard of weight and fineness, and to pay interest thereon from the date hereof at the rate of five per cent. per annum, until paid, such interest to be payable at such office or agency in like gold coin semi-annually on the first days of June and December in each year, but only upon presentation and surrender of the respective coupons for such interest hereto attached as they severally mature. All payments upon this bond, both of principal and interest, shall be made without deduction of any tax or taxes which the Subway Company may lawfully be required to pay thereon, or to retain therefrom under any present or future law of the United States, or of any State, county or municipality therein. This bond is one of a duly authorized issue of coupon bonds and registered bonds of the Subway Company, issued or to be issued pursuant to and secured ratably by and subject to an indenture dated June first 1908, duly executed by the Subway Company to the United States Mortgage and Trust Company, as Trustee, to which indenture reference is hereby made with the same effect in all respects as if the provisions of said indenture were herein fully set forth. All rights of action and otherwise of the holder hereof under or by reason of this bond are agreed by said holder to be expressly subject to the provisions of said indenture. No recourse shall be had for the payment of any part of the principal or interest of this bond against any incorporator, or any present or future stockholder, officer or director of the Subway Company, either directly or through the Subway Company, by virtue of any statute or by enforcement of any assessment or otherwise, any and all liability of the said incorporators and of the present and any future stockholders, directors and officers of the Subway Company being by the acceptance hereof and as a part of the consideration for the issue hereof, expressly waived and released. This bond is subject to redemption at the option and election of the Subway Company in the manner and upon the conditions and the publication of notice provided in said indenture on any day after the date hereof upon which an instalment of interest thereon shall mature and accrue, by the payment of the principal thereof and of all unpaid accrued interest thereon together with a premium of five per cent of such principal. This bond shall pass by delivery unless registered in the owner's name on the books of the Subway Company at its office or agency in the City of New York, such registry being noted on the bond by the bond registrar of the Subway Company, after which no transfer shall be valid unless made on said books in the manner prescribed in said indenture and similarly noted on the bond, but the same may be discharged from registry by being transferred in like manner to bearer, after which transferability by delivery shall be restored, but again from time to time this bond may be registered or transferred to bearer as above. Such registry, however, shall not affect the transferability of the coupons for the interest hereon by delivery merely and payment to bearer thereof shall discharge the Subway Company in respect of the interest therein mentioned, whether or not the bond shall have been registered. This bond with the coupons for all interest instalments which shall not have matured may also, as provided in said indenture, be exchanged for a registered bond without coupons in each case of the same series and for an equivalent principal sum on payment of the charges provided in said indenture. Neither this bond nor any coupon for interest thereon shall become or be valid or obligatory for any purpose until and unless the certificate endorsed hereon shall have been duly signed by the Trustee under said indenture. In Witness Whereof, said Chicago Subway Company has caused these presents to be signed by its President or one of its Vice Presidents, and its corporate seal to be hereunto affixed and attested by its Secretary or one of its Assistant Secretaries, and coupons for such interest bearing the engraved facsimile signature of its Treasurer to be attached hereto the first day of June, A.D. 1908.

Chicago Subway Company,
By
Vice-President.

Attest
Thos H Barrett
Assistant Secretary.

Chicago Subway Company bond.

Bruce G. Moffat Collection

board of directors. Railroads represented included the Erie, Santa Fe, Burlington, St. Paul, Southern Pacific and Union Pacific.

Following the acquisition of the two companies, the Subway Company formed a new subsidiary called the Chicago Warehouse & Terminal Company. Incorporated on December 21, 1904, the new company's purpose was to take responsibility for building, maintaining, and otherwise managing all tunnel connections to buildings and freight handling facilities located under private property. Freight customers now made arrangements for the movement of their goods with the Chicago Warehouse & Terminal Company, who in turn arranged with the Illinois Tunnel Company to actually move the cargo. The Illinois Tunnel Company retained ownership of all tunnels located under city streets and other public property and concentrated exclusively on operation of the train and telephone systems.

The next major step in the reorganization process did not occur until the June, 1907, meeting of the Illinois Tunnel Company's Board of Directors. At that

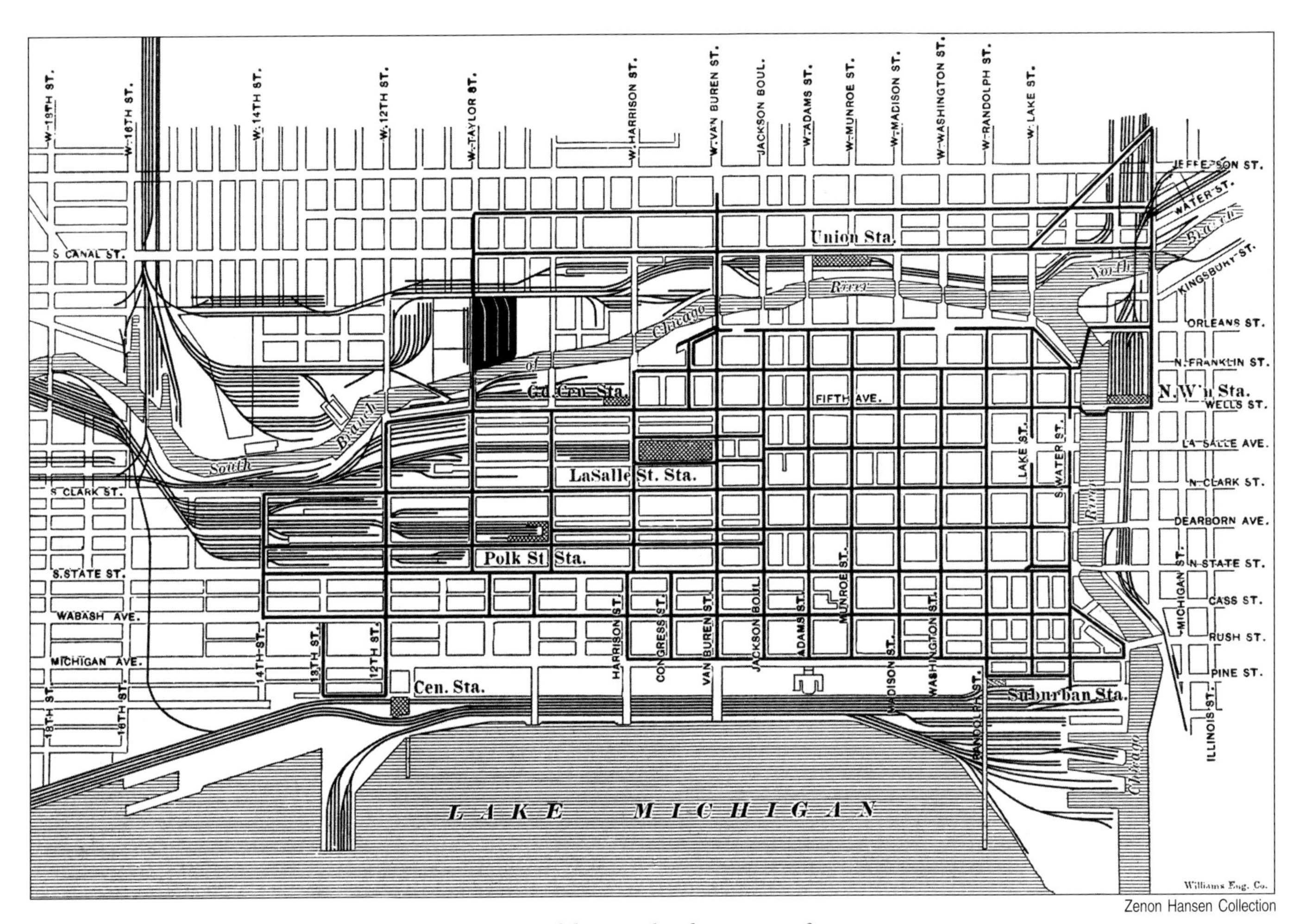

Zenon Hansen Collection

This contemporary map supposedly depicted the extent of the completed system as of 1905. The cartographer's information must have been based on an excess of company optimism, however, as the plat includes tunnels that were never constructed.

meeting, changes in the company's management and directorate solidified J. Ogden Armour's control of the organization. Wheeler resigned as Illinois Tunnel Company president and was replaced by Samuel McRoberts of Armour & Company, while W. J. C. Kenyon, who was manager of the Armour-controlled Omaha Stock Yards, was made general manager. Armour had himself elected a director, succeeding Valentine. Wheeler's "retirement" from the presidency was explained as being due to the fact that the company had passed the promotion stage, and that he thought it best to retire in favor of a man who was more suited for heading an operating railroad. Wheeler did retain a seat on the board, however.

On the financial side, the company's board was told that there were $17 million in outstanding bonds out of an authorized issue of $30 million and that the term on an outstanding $5.5 million loan had been extended to December 1, 1910. It was also announced that an additional $4 million loan had been secured for the completion of the tunnel system (what constituted "completion" was not defined). About 47 miles of tunnel had been completed and the bores were being extended at the rate of 1½ miles per month.

The Stock Yards Extension

In mid-1905, with the downtown portion of the tunnel system rapidly nearing completion, plans were drawn up for an extension of the bores southward to serve the industrial area along the South Branch of the Chicago River, with the Chicago Union Stock Yards being the ultimate goal.

The initial phase of this project called for the building of 20,460 feet of main tunnel plus whatever was required for sidings and building connections. This figure included 6,730 feet along Archer Avenue between Clark and Quarry Streets plus another 11,140 feet along various near south side streets including Canalport from Morgan to 18th, and along 18th to Clark.

Construction began on June 1, 1905, with the completion of a construction shaft in the west sidewalk of Armour Avenue (now Federal Street) just south of 17th Street. From this point tunnels were soon dug under Armour between 16th and 20th Streets, a distance of 1,513 feet, with shorter bores being completed on 18th, Dearborn, and Clark Streets, for a total of 2,763 feet.

The Illinois Tunnel Company, acting through its subsidiary Illinois Telephone Construction Company, was apparently following the same optimistic and ul-

A section of the ruptured water main.

Larry Best Collection

The torrent of water unleashed by the broken pipe completely demolished this nearby freight platform.

Larry Best Collection

timately costly construction policy it had used in building the rest of the system: building tunnels under every street regardless of business potential. The promoters believed that adjacent businesses would be eager to connect with the tunnel system so that they would have an alternative to the teamsters and their wagons.

The continuation of this somewhat naive policy, while risky in the heavily built-up downtown area with its concentration of large businesses, would have resulted in absolutely disastrous consequences had it been fully carried out in the area around 18th and Armour. According to the company's own 1905-vintage maps, this neighborhood was populated largely with small frame and modest-sized brick buildings having little or no traffic potential. A Wells Fargo & Company stable, a church and several grocery stores dotted the neighborhood (the area was also home to a number of bordellos). Clearly this area didn't warrant the construction of such an elaborate freight haulage system.

Why the company concentrated on building under area side streets instead of building a single line to contact the many large factories, wharves and freight terminals located nearby is a mystery at this late date. Also unclear is why the construction effort was not directed first toward connecting with the existing system at its southernmost point at 16th and State Streets.

A Calamity

The young enterprise was extremely proud of its safety record. In a 1906 promotional booklet issued by the Illinois Telephone Construction Company Jackson boasted that: "No deaths have occurred that can be attributed to tunnel construction…" This statement was evidently made in response to claims that tunneling operations placed workers' lives in danger. While technically true, Jackson's statement conveniently excluded an earlier incident that claimed the lives of two men.

During the early morning hours of Sunday, October 29, 1905, a 36" water main ruptured at 18th and Clark Streets. The fast moving water rapidly inundated

Larry Best Collection

Larry Best Collection

Settlement problems blamed on tunneling were corrected with the aid of timbers and screw jacks.

area buildings, forced a suspension of area streetcar service and even undermined nearby railroad tracks, derailing several standing cars in the process. Water soon reached the Illinois Telephone Construction Company's nearby shaft and, according to published news reports, managed to force its way through the air lock doors that protected the as-yet unfinished bores in that area.

Soon afterwards, tunnel construction superintendent Michael Barry arrived on the scene and decided he would descend into the tunnel to determine the extent of the water accumulation. Normally, the tunnel was kept filled with compressed air when workmen were present. This was done not only to minimize the potential of cave-ins during mining operations but also to prevent methane gas from seeping in from the surrounding clay. However since no work had occurred since the previous evening, and none was scheduled until late Sunday night, the air compressor had been turned off.

Neglecting to restart the air compressor and vent any gas accumulation, Barry entered the tunnel and was quickly asphyxiated. After waiting nearly a half-hour, his brother Patrick entered the shaft and was fatally overcome by the gas as well. Two other would-be rescuers nearly met with the same fate but were pulled to safety. It would be nearly a day before the tunnel could be ventilated sufficiently to allow searchers to recover the bodies of the Barry brothers. Following this misfortune all construction activity was abandoned and the shaft was sealed, forever ending all hopes of southward expansion.

In later years the company would attribute the abandonment of the project to a lack of funds. While money may have been in short supply, the death of two men, and the lack of any serious business potential, was enough to call an early end to the project.

That Sinking Feeling

Although great pains were taken to minimize disruptions to the construction work, the company's lawyers were kept busy defending claims that the tunneling work was causing streets and abutting buildings to settle more than was normal.

The problem apparently had become quite serious prompting the *Daily News* to run the following article in its June 13, 1905, editions:

> Huge buildings in the heart of Chicago that have withstood the gales and climatic changes of many years are beginning to settle and the impression is growing, in spite of many denials, that the underground operations of the Illinois Tunnel company are responsible. Not only has the Pullman building at Michigan avenue and Adams street shown the results of some unusual strain in the form of a fissure, from roof to basement on the Adams street side, but the wholesale house of Marshall Field & Co. has suffered as well.
>
> To-day huge timbers keep in place the great blocks of granite at the southwestern corner of the structure, Franklin and Quincy streets... Neither officials of Marshall Field & Co. nor of the tunnel company could be found to-day to discuss the situation.

Reports soon started appearing in the press about subsidence (sinking) problems in the streets. These depressions ranged in depth from a few inches to as much as one foot. Jackson defended his company's work, stating that the tunnels in question had been completed for three years and were solidly built. He attributed the problem to the construction of new skyscrapers, whose caissons were disturbing the subsoil.

Larry Best Collection

To make its point that most street settlement problems were actually caused by the construction of large buildings and not the tunnels, the company photographer recorded this depression that developed in front of a new building at the corner of Jackson and State.

The city's public works commissioner was of a similar mind, stating that the construction of foundations for big buildings tended to draw water away from neighboring buildings causing the earth to shrink.

On June 30, the *Daily News* carried this probable explanation:

> The blue clay formation contains pockets of sand and pockets of water and when excavation is made some of this sand or water escapes from an opened pocket, and seeps into the excavation, from which it is pumped later. The seam of sand runs out and leaves the seam which must necessarily be filled sooner or later.

The city's superintendent of streets had a much simpler theory. He attributed the subsidence to huge sewer rats eating through the curb walls of older buildings.

The public works commissioner hired three engineers to investigate. Although the Illinois Tunnel Company denied any responsibility for the situation, they agreed to pay for the study.

In late August the engineers presented their findings. There was good news and bad news for both sides. The engineers reported that they were unable to find any evidence that the settling had resulted from the construction of the main tunnels that were built under pressure. However, when it came to the construction of connections to buildings and bypasses built without the use of air pressure, the commissioners laid the blame at the feet of the company. The engineers also reserved some of the blame for the city, saying:

> The control and supervision by the city of the work of building these tunnels has been lax when compared with similar work done for private individuals or corporations. Ordinary practice demands definite official plans and specifications and a sufficient force to supervise the work properly at all points and at all times; while in this case there were no definite plans and specifications and there was but one city representative in supervision of the entire work.
>
> We find the city records of the work incomplete. For example, we have been unable to find any written specifications either of the cement or the concrete. We also find the city has no official survey showing the exact location of the tunnels...

Bruce G. Moffat Collection

Inspection by city officials on March 13, 1906. In foreground, from left: Mayor Edward F. Dunne (standing next to Jeffrey locomotive 119), police chief John M. Collins, George W. Jackson, streets superintendent Michael J. Doherty, and city electrician William Carroll.

The engineers also investigated the financial condition of the company, stating in their report that a review conducted by a financial expert that they had hired determined that the company was overcapitalized "to the extent of probably 85 percent of its total issue of $57 million in stock and bonds." It was estimated that the cost of building the tunnels and acquiring property was not more than $8 million.

When questioned by reporters, Wheeler expressed surprise saying: "This talk of overcapitalization is without foundation. Why, we have distributed over $12 million in salaries and wages since we began the construction of the tunnels, and that is only a fraction of the total cost." Wheeler also maintained that settlement damage caused by the construction was insignificant, being less than $5,000. The city's Board of Review, which reviewed assessments for taxation purposes, valued the tunnel property at only a half million dollars. The city's corporation counsel said that the Board of Review had failed to assess the tunnel as a property and in his opinion, the company should pay taxes based on a valuation of $20 million. Clearly the cost and value of the tunnels and related assets depended on who was asked.

City engineer John Erickson took vigorous exception to the engineering-related aspects of the findings, saying that portions of tunnels constructed under public property for which permits could not be found had in fact been constructed under proper authority (most of the missing permits were eventually found). Further, he stated that he had a survey prepared of the tunnels' exact location. The surveyors' measurements were then verified by driving pipes into the ground through the center of the streets and through the roofs of the tunnels.

Another charge, that some of the tunnels were

Bruce G. Moffat Collection

W. G. Collins served as president during the 1905-1906 period.

built too close to the surface in violation of the company's franchise, was eventually refuted when further research showed that these tunnels were built before the 1903 franchise revision which had added the minimum depth requirement.

It took months for the tempest to die down. The end result was that the company apparently escaped most of the blame but was forced to take very elaborate precautions to prevent settlement problems when penetrating building foundations and walls to make the necessary connections with the main tunnels.

Revenue Operations Begin

During the first half of 1906, the *Railroad Gazette* carried a series of articles describing the construction of the tunnels and predicted that, with the completion of connections to the city's numerous railroad freight houses, the system would be ready for service on June 1. It was also predicted that within three months freight movements would reach 30,000 tons per day. These pronouncements were in addition to the by-now usual claims that surface congestion would be dramatically reduced as more shippers discovered the benefits of moving their goods by tunnel instead of relying on the teamsters and their slow horse-drawn wagons.

During this time, some coal shipments were handled between the just-completed underground coal receiving chute serviced by the Chicago & Alton Railroad and the boiler room of the First National Bank Building in the heart of the Loop. These movements were rather limited, and were of an experimental nature. Their primary purpose was to serve as a means to test the system and gain practical experience in its operation.

June 1 came and went without any formal inauguration of operations. By August 15, however, whatever problems had delayed the start-up had evidently been overcome and revenue service was officially begun. Approximately 45 miles of tunnel had been completed and were available for use in an area that encompassed about two square miles.

Though the event went largely unnoticed, a reporter from the *Evening Post* was on hand for the event:

> This morning at 11 o'clock the first trainload of merchandise traveled over the tracks in the network of tunnels underlying Chicago's streets.
>
> "This move inaugurates," said George W. Jackson, chief engineer of the tunnel company, "a system of freight handling that will relieve the streets of 20,000 tons of merchandise a month. It will revolutionize traffic in Chicago; will make teamsters' strikes impracticable; will give the passenger traffic of the downtown streets a chance to make some headway."

That first train consisted of eleven freight cars loaded with general freight and began at the Erie Railroad freight house located at 14th and Clark Streets. Going along on the initial run were Illinois Tunnel Company President W. G. Collins, George W. Jackson, and several freight traffic representatives of the Erie. Four cars were delivered to the freight house of the Chicago Milwaukee & St. Paul, two to Marshall Field's department store on State Street, four to the Chicago Great Western's freight house, and one to the Monarch Refrigerator Company at Cass and Michigan Streets (now Wabash Avenue and Hubbard Street).

Like its telephone counterpart, the railway seemed to generate more losses than profits. The actual amount of business handled was far below the level predicted by the *Railroad Gazette* and the more optimistic 50,000 tons per day forecast by management at the 1904 banquet. Losses from operations were covered by capital funds, resulting in a series of loans being secured in an attempt to keep the fledgling venture alive.

August 15, 1906: Opening Day

Never one to overlook a promotional opportunity, General Manager and Chief Engineer George W. Jackson was on hand to escort a group of railroad executives on a tour of the company's facilities on opening day.

Chicago Historical Society (DN-3993)

The first official carload of freight to be handled through the freight tunnel has just been received at the Milwaukee Road's freight house.

Chicago Historical Society (ICHi-23288)

Larry Best Collection

An overall view of the Milwaukee Road freight house on opening day.

At this time the elevators lacked even safety gates or doors, subjecting employees to great risk should they lose their balance.

Larry Best Collection

A group of Milwaukee Road officials pose with two carloads of merchandise that would barely fill a corner of a standard size railroad freight car.

Larry Best Collection

Municipal Reference Library

Excavation work for a new building on the southwest corner of Adams and State streets appears to be well along in this circa 1914 view. Note the 'x' marking the shaft was used to drop construction spoil into waiting tunnel cars. The streetcar in the background is on Adams.

Receivership

The succeeding years were not good ones for the Illinois Tunnel Company. Losses became the order of the day as expenses outpaced revenues. Publicly however, the company's officials projected a positive image, trumpeting the system's ever-growing reach. By early 1907, the system had grown to 46 miles of tunnel having 41.53 miles of permanent track and 39.67 miles of overhead trolley installed. The tunnels now extended north to Erie Street, west to Halsted Street, and south to 16th Street. Connections had been made with 20 major Loop buildings, nine principal railway freight houses and two coal transfer facilities. The rail fleet had grown to include 65 locomotives and 590 cars. Little mention was made of the telephone side of the operation which was being allowed to languish so that the railway could be built.

Despite these glowing statistics, the vast majority of the Loop's commercial, industrial and office buildings remained disconnected. Many were simply too small to reap any significant benefits from receiving coal and other supplies by tunnel. In many cases the high cost to reconstruct many of these buildings with deeper basements or elevators made use of the tunnels impractical. This translated into diminished profit potential.

This situation had not escaped the notice of major tunnel backer Edward H. Harriman who was quoted in a March 14, 1907, *Daily News* article as saying:

> Those tunnels ought to do well if there were cooperation on the part of the business people of Chicago. But so far as I can understand, it is hard work to get the merchants to have their places connected up. I don't think the people quite understand how beneficial those tunnels could be made.
>
> If we were given the privilege of carrying light, heat and power, as well as telephone wires, we could do a great deal toward abolishing smoke. Chicago out to be one of the brightest and cleanest cities in the country. If we were given the right to do what I suggest it would result in removing a great many of the small smoke-making furnaces.

(Use of the tunnels for steam heating purposes was three decades away while the transmission of electric power would not happen until well after the 1959 abandonment of the railway.)

Mr. Harriman's remarks can best be described as an understatement. In fact, the operating results for the transportation portion of the business failed to approach the preliminary estimates made by the promoters. Whereas they had predicted handling 10,000 tons of cargo daily, the actual amount proved to be only 2,000 tons. Similarly, less than 100 tons of coal was being handled per day instead of the 500 tons predicted. Construction spoil car loadings were far below expectations.

Operating costs also exceeded earlier estimates.

Bruce G. Moffat Collection

A bottom-to-top view of a typical elevator shaft with one of the inspection cars loaded aboard.

It gradually became apparent that the tunnels were simply too small, thereby precluding the handling of many large and heavy articles such as machinery and even furniture. These items had to be handled by the teamsters who would not accept them without being allowed to handle some of the small freight that could have been routed through the tunnel. As a result, the company was deprived of considerable traffic.

On December 1, 1908, the Illinois Tunnel Company moved closer to bankruptcy when it defaulted on interest payments covering more than $13 million worth of bonds. In 1909, the Corporation Trust Company of New Jersey, holder of $1.1 million in Tunnel Company notes and $2.7 million in Warehouse

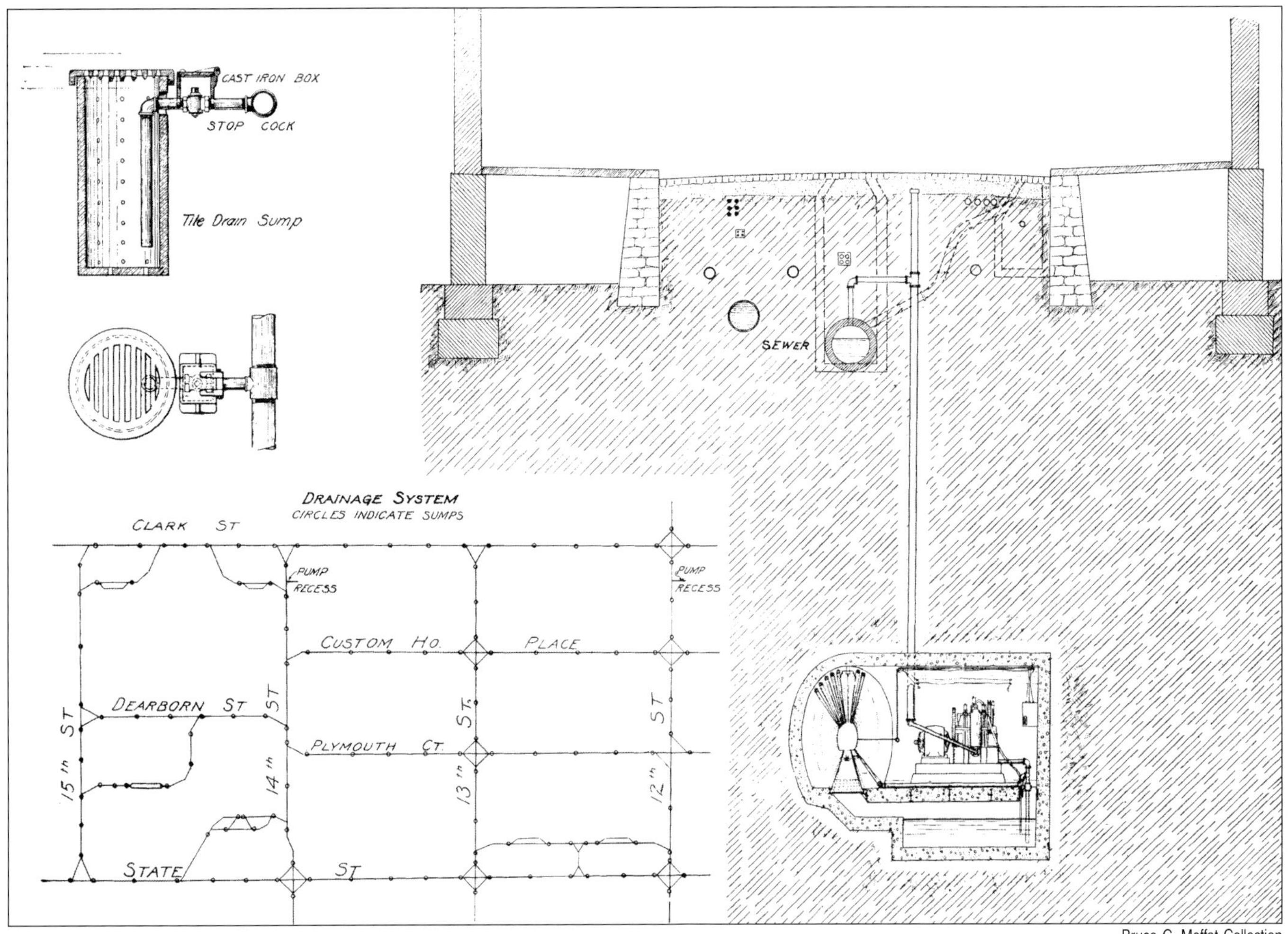

Bruce G. Moffat Collection

The drainage system needed to keep the tunnels reasonably free of water was quite elaborate as illustrated in this drawing that covers just a small portion of the network south of the Loop.

Company notes filed suit in U.S. District Court in Chicago to collect from both companies. The court returned judgments in favor of the Corporation Trust Company.

To further complicate matters, the passing of Harriman on September 9, 1909, caused the executors of his estate to announce their intention to terminate their association with the tunnel venture. This spread consternation among the remaining investors. Divestiture was averted when Mary Harriman consented to retain all of her late husband's holdings on the condition that J. Ogden Armour take over the active management of the syndicate.

The judgments obtained by the Corporation Trust Company, combined with Harriman's death, alarmed the unsecured creditors. These creditors quickly petitioned the U.S. District Court to place the Illinois Tunnel Company and the Chicago Warehouse & Terminal Company in receivership. On December 1, 1909, Judge Kohlsaat appointed David R. Forgan and Charles G. Dawes as receivers for the Illinois Tunnel Company, and Edwin A. Potter, receiver for the Chicago Warehouse & Terminal Company.

Immediately after taking over the properties the receivers prepared inventories of the properties under their supervision. That portion related to the Illinois Tunnel Company's railway operations is summarized below:

Completed tunnels (incl. intersections and railway switches)	58.3 miles
Railway tracks in tunnels	57.6 miles
Trolley system in tunnels	57.6 miles
Feed cables in tunnels	24,800 feet
Drainage pumps	73
<u>Cars:</u>	
Flat	50
Box (spoil/ash)	195
Registered mail	15
Inspection	5
Coal	235
Freight (merchandise)	2,001
Wrecking	1
Total:	2,502
Electric locomotives	125

Jeffrey Division, Dresser Industries

Ordered on behalf of the Illinois Tunnel Company by the Illinois Telephone Construction Company in early 1905, this Jeffrey-built locomotive (serial # 929) was assigned fleet number 112.

Receiver Potter reported that the Chicago Warehouse & Terminal Company had about 50 contracts covering tunnel connections of which 26 were with railroads (these 26 connections used 60 elevators). The aggregate length of the trackage used in these connections (and technically under the CW&T's ownership) was 33,194 feet.

Throughout the period of the receivership efforts were made to adjust finances, alter practices, and adopt new procedures to place the two companies on a profitable operating basis. To further this end, Illinois Tunnel receivers Forgan and Dawes appointed C.O. Frisbee, general traffic manager for Armour & Company, to the post of receivers' agent to manage the railway operations, and Joseph Harris, president of the Automatic Electric Company, to manage the telephone system.

Under receivership, conditions did improve somewhat. On July 1, 1911, Frisbee reported that the freight system was earning a modest profit of about $7,000 per month. His report also included the following statistics:

Tunnel Equipment:	
Locomotives in service	117
Locomotives rented and out of service	17
Coal, freight and other cars	3,001
Track Mileage and Connections:	
Freight house connections	6.047 miles
Business and coal connections	2.970 miles
Mail Connections	1.016 miles
Street and intersections	45.494 miles
River tunnels	3.315 miles
Total tunnels:	58.842 miles
Railroad (elevator) shafts	58
Commercial shafts	33
Total shafts:	91

Bruce G. Moffat Collection

Once on the property, the locomotives quickly took on a well-worn appearance as evidenced by this pre-World War I view of a Jeffrey maneuvering an ash car. Note the ornate headlight cover.

How the company effectively managed this large fleet, given the absence of any dedicated car storage or marshaling facility, is left open to speculation.

The receivers also purchased 500 new freight cars and constructed several new building connections. In addition, funds were expended to rehabilitate and paint the rolling stock, settle unpaid freight bills and pay back taxes. Unfortunately, these promising developments were offset by the receivers' decision to request prospective customers to pay half of the construction cost for a connection. This resulted from a study that found that a typical building connection would not be profitable unless its daily volume was at least 40 tons. The result was the loss of many potential patrons whose business might have been profitable, but were unable - or unwilling - to assume any part of the construction cost.

The telephone portion of the business, however, failed to break even, resulting in overall operating losses being reported for the system as a whole. These losses were covered through the use of funds that had been earmarked for equipment depreciation and other expenses.

Bruce G. Moffat Collection

Isssued by a warehouse company, this promotional paperweight noted their tunnel connection.

4 Dial Telephone Service Comes to Chicago

Although relegated to a secondary role following the enactment of the amended franchise ordinance on July 15, 1903, the telephone system nonetheless led an interesting existence.

The Illinois Telephone & Telegraph's amended franchise from the city required that it provide service to 20,000 subscribers within five years, or risk the forfeiture of all rights, as well as title to its telephone plant and equipment, to the city. When the Illinois Tunnel Company succeeded the IT&T, the new company inherited this requirement. Although Wheeler professed his commitment to install 250,000 phones and provide service to the entire city, the reality was far different. The *Tribune* probably best summed-up the company's true priorities when it stated in its October 30, 1903, editions that "Telephones will be merely a

Larry Best Collection

The original telephone exchange was housed in this building on Fifth Avenue (Wells Street).

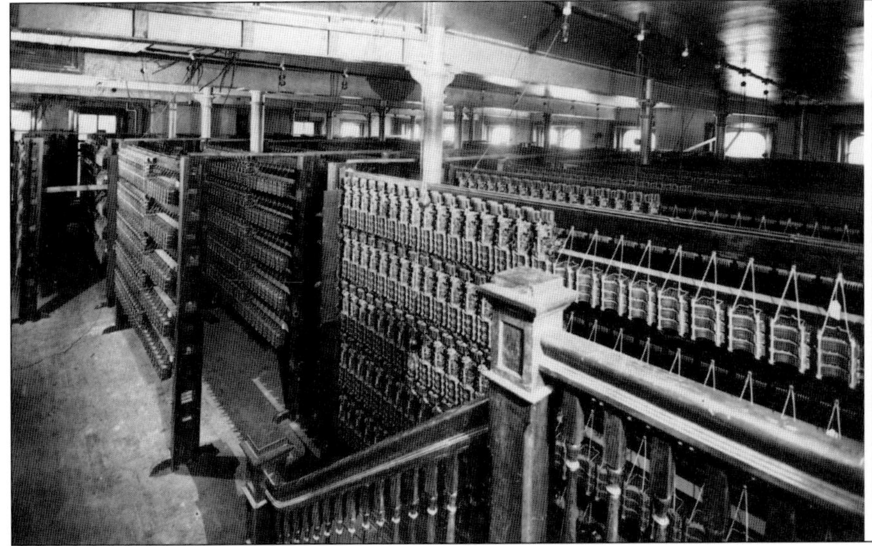

Bruce G. Moffat Collection

The building's upper floors housed row after row of these formidable looking switch boards.

Looking north on Fifth Avenue (Wells Street) from Monroe in 1904. Although the telephone cable conduits are suspended above the Morgan-equipped third rail track, space has been left for the future installation of trolley wire.

Chicago Historical Society (ICHi-31234)

side issue with the company. There is space in the conduits for the telephone wires and that is said to be the reason the automatic service is being promoted. The real object is the transportation of freight."

Telephone service was inaugurated in a low-key manner on October 15, 1903, when the first exchange was placed in service at 177 Fifth Avenue (now 105 South Wells Street). The press paid little attention to this upstart competitor to the long-established Chicago Telephone Company until February 8, 1904, when the successor Illinois Tunnel Company arranged for the *Journal* and several other Chicago newspapers to publish a sensationally-worded promotional "news article" that it had commissioned. The attention-grabbing headline proclaimed "CHICAGO'S NEW MODERN AUTOMATIC SECRET SERVICE TELEPHONE EXCHANGE." The word "secret" referred not to its heretofore obscure existence, but rather to the fact that, by being able to simply dial the desired number, a telephone subscriber did not have to tell an operator the number or name of the party to be contacted. Besides, an automatic switching system could not eavesdrop on conversations and spread gossip.

By the time the article appeared, the company claimed that more than 6,000 telephones were in use. In August 1904, the *Journal* reported that 8,000 instruments were in use and were all situated in the city's central business district bounded by Lake Michigan, the Chicago River and 12th Street (Roosevelt Road). Since the preponderance of telephone traffic at this time was between businesses, the company showed little interest in soliciting residential customers. This also allowed the company to confine its capital investment to just the one switching plant which had a reported capacity of 10,000 instruments.

In an effort to attract as many subscribers as possible, free service was given from October 1, 1904, until April 1, 1905. However once the company began to charge for its service it experienced difficulty retaining subscribers. Many subscribers apparently liked the service, but only as long as it was free.

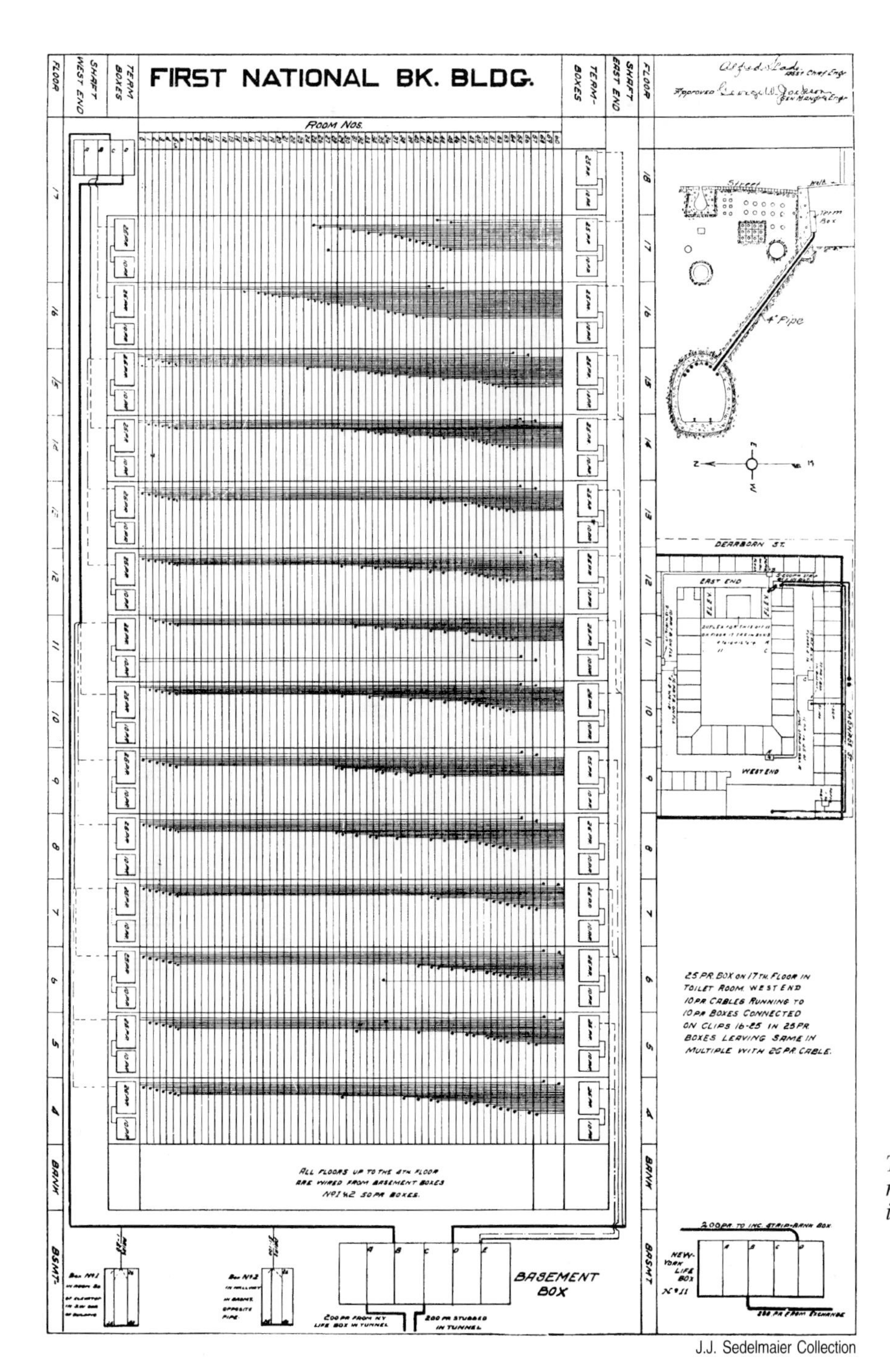

J.J. Sedelmaier Collection

Telephone wiring diagram showing the method used to layout telephone lines in large downtown office buildings.

Telephone Installation Effort Halted

On February 1, 1905, the *Tribune* carried this announcement:

> It is announced that the Chicago Subway Company [the holding company] has discontinued installing its automatic telephones. There are nearly 10,000 instruments in service, or practically a sufficient number to conform to the terms of the franchise. No more work, it is said, will be done on the phone service until the tunnel work is completed. After that steps will be taken to extend the wires over an enlarged area."

Whether this announcement stemmed from a shortage of working capital, or was merely an early sign of disinterest on the part of management, is unclear. It is also unclear how the company's assertion that having "nearly" 10,000 telephones in service could be deemed "practically sufficient" to conform with the terms of the 1903 franchise that required that 20,000 instruments be in service within five years.

What is certain is that while the company was investing very heavy sums on constructing the railway system, no serious effort was being made to retain and attract telephone subscribers. In fact, one estimate placed the number of actual subscribers in 1905 at just 3,500.

The Cheese Store War

In setting up its telephone system, the Illinois Telephone & Telegraph Company arranged to lease space in various building basements for equipment storage and to house telephone line connection boards. In January 1905, the company's construction forces managed to arouse the ire of John G. Neumeister, the proprietor of a Loop cheese shop. The cheese merchant had earlier agreed to lease to the company a portion of his basement.

Unable to come to terms with the company concerning the unauthorized consumption of "his" electricity by the company's workmen, Neumeister refused to pay his bill. A former military veteran of unknown rank, the self-styled "general" declared war against the IT&T when the telephone company's "forces" attempted to retaliate by removing his telephone. The ensuing escapade, which capped a six-month feud between the parties, was dutifully recorded by a reporter for the *American* and appeared in print on January 21, 1905. With tongue planted firmly in cheek, the story was presented as a collection of wartime news dispatches:

Automatic Electric Company

Neumeister's telephone was a typical Strowger-type wall telephone of the period.

HEADQUARTERS
GENERAL NEUMEISTER'S ARMY,
FORT LIMBURGER,
154 SOUTH WATER STREET.

Bulletin (9:30 a.m.) – General John G. Neumeister, commanding the forces at Fort Limburger, is momentarily expecting an attack by the allied forces of the Illinois Telephone & Telegraph Company and the Illinois Tunnel Company. They intend to remove his automatic telephone. General Neumeister says Fort Limburger may be his grave, but the telephone will remain until that event is recorded in history.

Bulletin (9:50 a.m.) – The armament of the fort consists of one brass cannon, brought from Cuba by General Neumeister after the Spanish war; one Winchester rifle, operated by the General in person; four revolvers, distributed among cheese salesmen and girl stenographers, and the trusty saber worn by General Neumeister when he was a colonel on Governor Altgeld's staff. The cheese is held in reserve and will not be used except as a last resort.

Bulletin (10:04 a.m.) – Your correspondent, at great personal risk, has just secured a statement from General Neumeister, "the lion of South Water street," on the casus belli. He said: "The Illinois Telephone Company wanted to use my basement for a switchboard for this district. That was all right until I discovered their workmen were using my electric light current while at work and my lighting bills were going up. Protests were useless and I finally drove the men out and refused further access to my property. Then the company cut off my telephone service. My contract provides for a thirty days' notice in writing before the instrument can be removed. I have not received such notice and the company intends sending three men to take away the telephone to-day. They will not get it."

Bulletin (10:21 a.m.) – General Neumeister has just cleaned and oiled his Winchester again and now wears a look of determination.

Bulletin (10:30 a.m.) – Scouts from Dearborn street report by Bell telephone that the enemy's repair wagon has been operating in the loop district. Excitement in Fort Limburger is intense. "If we are driven back," said General

"HOLD THE FORT," LOUDLY CRIES CHEESE MERCHANT NEUMEISTER

Proposed Raid by Telephone Company to Be Met With Brass Cannon, Winchester Rifle and Other War Implements.

Larry Best Collection

"General" Neumeister is poised to defend his telephone (visible at top right).

Neumeister, "we will use our handcase grenades with smelling – I mean telling effect."

Bulletin (10:38 a.m.) – Brass cannon has been polished anew by porter, and now looks more formidable than ever. South Water street is agog with excitement. All is tension. The blow may fall at any moment.

Bulletin (10:42 a.m.) – General Neumeister has just issued the following statement: "We have nothing to arbitrate.

Apparently, the *American's* intrepid "war correspondent" left the "front" before the enemy's assault because no further dispatches were received that day. However, in the next day's editions, it was reported that Neumeister had turned to the courts in an effort to have his telephone service restored. Unfortunately, the judge's decision escaped media coverage.

Seeing Double

Chicago-based catalog giant Sears Roebuck & Co. utilized the Strowger system for in-house communications at its corporate headquarters and warehouse/catalog sales complex on Homan Avenue. Because of the large number of telephones required, Richard Sears' company had its own PBX (Private Branch Exchange) which was supplied by the Illinois Tunnel Company. The system was considered so modern at the time that it was featured in a set of steroscope images that described Sears' operations.

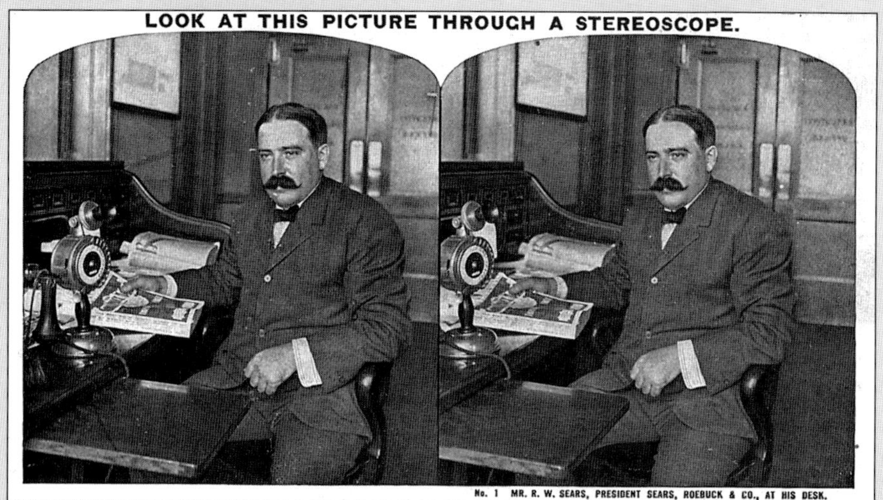

Bruce G. Moffat Collection

Company founder and president Richard Sears at his desk with a Strowger phone. Note how the right and left images, although identical, are cropped differently to enhance the three-dimensional effect.

Bruce G. Moffat Collection

The PBX "switchboard." The card's reverse side included the following explanation.

No. 15. AUTOMATIC TELEPHONE SWITCHBOARD.
Sears, Roebuck & Co., Chicago, Ill.

This view of an interior in our great mercantile institution represents what is undoubtedly one of the great inventions of this electrical age. It is located in the basement of the Administration Building and is our automatic telephone switchboard. We have installed automatic telephones throughout our entire establishment for interdepartment communication. We therefore own our own telephone plant as used by department managers and their assistants. The great advantage of this system lies in the fact that the switchboard is purely automatic, no switchboard operators being required. You will notice in the foreground, resting on the bell box, what looks like an ordinary telephone. In the center of this instrument you will observe a dial with notches around the right half; these notches are numbered from one to ten. This dial revolves to the left, and to get a given department, say No. 123, the receiver is taken off the hook, the index finger inserted in notch number one and the dial pulled around to the left until the finger strikes the index just to the left of the lower center of the dial, when the dial is released and permitted to return to its original position; then the finger is inserted in notch number two and the operation repeated; then the finger is inserted in notch number three and the dial turned to the left as before. This has registered in the automatic switchboard the connection for phone No. 123, and when you press a button in the base of the instrument it rings the bell in department 123, just the same as it would be rung by a girl calling from the ordinary switchboard. The great advantage in this style of telephone is that communications pass from one department to another in absolute confidence. It is not possible for anyone else to hear a conversation passing over the wire. We have 325 automatic telephones installed in this plant. In addition to this automatic telephone system we also employ the services of the Chicago Telephone Co. for long distance purposes, as illustrated and explained in another view.

Enterprising Customers

One group of customers that embraced the new technology and had no qualms about paying their bills were the city's local bookmakers or "bookies." For quite some time they had been using the lines of the Chicago Telephone Company to receive wagers on horse races from their "clients."

The police attempted to keep this illegal activity in check by enlisting the aid of telephone operators who were known to regularly eavesdrop on conversations routed across their switchboards. Intent on preserving their very lucrative business, the bookies responded by bribing operators with flowers, candy, and similar tokens to ensure their silence so that they could continue to ply their illegal trade. The automatic telephone system's technological advantage of guaranteeing the utmost confidentiality in the completion of calls fit their needs perfectly and they promptly switched over. The fact that the company did not keep records of originating calls added an additional layer of assurance.

Attempt Made to Sell System

In 1907, an attempt was made to sell the loss-plagued operation to the Chicago Telephone Company who would have been only too happy to oblige. In a lengthy article about the Chicago situation, *Telephony* magazine noted that "the Tunnel Company did not realize what a good business it was trying to throw away. But Chicago was alive to the risk of a telephone monopoly."

Approval of the sale rested with the City Council which began a very protracted round of hearings on the subject. Both the Illinois Tunnel Company and the Chicago Telephone Company presented fiscally based arguments for proceeding with the sale. Arguing against the sale were the influential City Club of Chicago and the International Independent Telephone Association, a trade organization backed by the Automatic Electric Company. Not surprisingly they both agreed that Chicago needed to retain its competitive telephone environment.

Speaking on behalf of the City Club at one hearing, prominent local traction attorney Walter L. Fisher stated: "We are very apprehensive that this committee, perhaps through mere inadvertence, may be led into taking some action which will vitally affect the question of rates in this city. If your committee adopts the policy of permitting a sale to the Chicago Telephone Company it will be because this committee has decided that the principle of regulated monopoly in the telephone business is the correct principle. It cannot be justified upon any other theory." Ultimately, the City Council agreed with Fisher and voted to deny the sale.

Left with a largely idle investment on its hands, and with its railway operations generating insufficient income to meet expenses, the Tunnel Company was forced to consider seriously the question of whether to extend the reach of its telephone system or sell it outright.

Chicago Historical Society (DN-5301)

A 1907 view looking west on Monroe at LaSalle showing the arrangement of the telephone conduits and trolley wire.

The need to address this situation gained some urgency when, on February 1, 1909, the City Council amended the company's oft-modified franchise to change the deadline for having 20,000 bonafide subscribers from July 15, 1908, to October 8, 1909. The legislation also stated that if at any time the number of subscribers fell below that number the company would be required to forfeit its rights to operate the telephone system and turn over ownership of all telephone-associated equipment to the city. Apparently the solicitation of subscribers was not going well at all, because on June 28, the aldermen extended the deadline to June 1, 1911.

The company's long-term solvency came into question following the death of backer E. H. Harriman on September 9, 1909. A short time later, the executors of his estate announced their intention to withdraw from the tunnel and telephone enterprises. As mentioned earlier, Harriman's widow, Mary, had no interest in attempting to manage the syndicate's vast holdings herself. Fortunately, to the great relief of the other investors, J. Ogden Armour agreed to take over active management of the syndicate's portfolio.

Harriman's death, combined with two judgments levied against the Illinois Tunnel Company and the Chicago & Warehouse & Terminal Company for more than $4 million, caused alarm among the unsecured creditors. Hoping to preserve their investment, an application was made to the U.S. District Court to place both the Illinois Tunnel Company and the Chicago Warehouse & Terminal Company in receivership. The receiverships became effective on December 1, 1909.

Although characterized in the financial press as "friendly" receiverships, the court-appointed receivers and the bond-holders reorganization committee were obliged to determine whether the remedy should be partial or total liquidation of the railway or telephone operations, or expansion of one or both components in the hopes of achieving greater income.

Whatever the underlying concerns were about the potential economic viability of the telephone operation, they must have been largely resolved, because in early 1910, the reorganization committee and receivers of the Illinois Tunnel Company approved a plan to renovate the telephone system. Approved by the judge overseeing the company's reorganization on March 16, 1910, the $2.5 million plan provided for the installation of the 20,000 telephones by June 1, 1911. The plan also called for expansion of the "independent telephone system" (a phrase used to identify any "non-Bell" telephone company) to serve all parts of the city as quickly as possible. Additionally, plans were being made to connect with lines of the various other small independent companies at the city limits. Receiver's certificates were sold to raise the needed cash. Ultimately, the bankruptcy court authorized the sale of $4.5 million in certificates to cover the cost of the work.

Prominent in the negotiations which surrounded the receivership and the decision to expand the telephone system was Joseph Harris, president of the Automatic Electric Company, whose company's fortunes were tied to the continued expansion of the independent telephone company movement and its heavy reliance on Automatic Electric Company-manufactured equipment. Perhaps not surprisingly, the Tunnel Company awarded its lucrative telephone installation contract to Harris, who then established the Subway Telephone Construction Company. (Harris was also made General Manager of the Tunnel Company's telephone activities by the receivers.)

In a March 19, 1910 interview, Harris outlined his plans:

> We will install a complete new automatic telephone plant of the latest and most approved type of apparatus manufactured by the Automatic Electric Company of this city with 20,000 main line telephones to be in operation by June 1, 1911, and as many more telephones thereafter as the public requires. The hard and fast requirement of 20,000 telephones within the time fixed is made to conform to the conditions of the ordinance under which the plant will be installed. The apparatus furnished will be of the type which is now in service in approximately one hundred cities and towns in the United States as well as Canada...
>
> With the first installation comprehensive long-distance connections will be made, giving connections to Chicago business houses with more than 1,300,000 Independent telephones within five hundred miles of Chicago. These embrace fully one-half of all the business telephones of both Independent and Bell lines within that territory.
>
> The contracts have been duly executed, all final details are arranged and construction work will begin immediately.

Harris' pledge to install new equipment was more than just public relations hype. The existing switching equipment was considered obsolete and was not well adapted for use in a system as large as the one now proposed. Essentially, Harris' task was to install and promote a completely new automatic telephone system.

As far as service to the city's Loop district was concerned, there was little that Harris had to do except convince people to sign up:

> Ordinarily, this would seem like a large undertaking within the time fixed for the first installation to be completed, but owing to the fact that approximately 70 miles of tunnels, covering practically all the business district ... are open; that a vast amount of cable is now in the tunnels in good condition which will be available for new installation, and that most of the large buildings in the downtown district are wired for this system, there is a comparatively small amount of work to do and the contract provisions can easily be met.

From his comments, it appears that he had little interest in soliciting business much beyond the industrial zone that ringed the Loop district. Harris evidently believed that this area had significant potential, or at least enough to satisfy the 20,000 subscriber requirement. Wiring homes in the city's far-flung (and less profitable) residential areas still remained a lesser priority although a few areas were eventually wired-up.

On April 1, 1910, the Subway Telephone Construction Company launched an aggressive advertising campaign to acquaint Chicagoans with the advantages of "Automatic Telephone Service." As many as 250 solicitors were hired to induce businesses and private citizens alike to sign up.

Eye-catching full-page advertisements placed in the Chicago dailies proclaimed: "The Automatic apparatus to be used, is so simple that a child can operate it without error." A writer for one telephone

THE CHICAGO DAILY TRIBUNE: TUESDAY, MAY 2, 1911.

Most Important Telephone Announcement

In Chicago's History

The Illinois Tunnel Company has just completed arrangements to connect the independent telephone users in the territory tributary to Chicago with its automatic telephone system in this city. This assures a city-wide automatic telephone service—17,000 telephones being already installed and more than 20,000 will be in service by June 1st of this year. This is the quickest initial installation of this number of telephones ever made.

The conduit work is already completed with facilities to accommodate one hundred and fifty thousand subscribers and Chicago will ultimately have five hundred thousand automatic telephones.

Automatic telephones mean vastly superior service at much lower cost; unlimited number of calls and elimination of all troubles incident to the manual operation of a telephone system.

Flat rate, unlimited automatic service, means all the calls which can be put over a telephone circuit without increasing the cost either to the subscriber or the company.

Restricted or measured service costs the user additional money every time a call is made.

With our unlimited automatic service the cost does not increase with the number of calls.

Automatic telephony is a radical improvement in the telephone art.

The perfection of a faultless machine which dispenses with the human operator, has made possible a telephone service which costs no more to the user and no more to furnish by the company, for an unlimited number of calls than it does for one call.

The people of Chicago are each year paying millions of dollars more for telephone service manually operated, than they will be required to pay for a vastly superior service, automatically operated.

With this service you are not penalized for getting the maximum benefit from the use of your telephone.

The automatic telephone means greater efficiency than can possibly be had by the old manual method.

It means the elimination of delays; wrong connections; disconnects during conversation. Think how often these and other annoyances occur to you and what a relief it will be to be free from them.

Automatic telephone service is equally efficient days, nights, Sundays and holidays.

If you are a present or prospective telephone user, if you have suffered from inefficient telephone service, interest yourself in the Automatic.

The saving in time, money, annoyances, temper, wear and tear of human nerves is enormous.

Twenty-Five Thousand Dollars in Prizes

Every Man, Woman, Boy and Girl in Chicago Is Invited to Participate

We want you to familiarize yourself with this proposition and carry the news into every home and place of business in the city and are ready to pay you liberally for so doing.

We will, subject to the conditions below, pay to the persons securing service contracts the following prizes and compensation:

To the person securing the LARGEST number of service contracts on or before July 1, 1911 - - first prize, ONE THOUSAND DOLLARS

To the person securing the NEXT LARGEST number of service contracts on or before July 1, 1911 - second prize, FIVE HUNDRED DOLLARS

To the FOUR persons securing the four next largest number of service contracts on or before July 1, 1911 - TWO HUNDRED AND FIFTY DOLLARS, EACH

To the SIX persons securing the next largest number of service contracts on or before July 1, 1911 - ONE HUNDRED DOLLARS, EACH

To the TWENTY persons securing the next largest number of service contracts on or before July 1, 1911 - - - - FIFTY DOLLARS, EACH

To the THIRTY-SIX persons securing the next largest number of service contracts on or before July 1, 1911, TWENTY-FIVE DOLLARS, EACH

To the ONE HUNDRED persons securing the next largest number of service contracts on or before July 1, 1911 - - TEN DOLLARS, EACH

In addition to the foregoing, and subject to the conditions hereof, we will pay FIFTY CENTS for each service contract, regardless of whether or not you are one of the prize winners. All contracts taken hereunder, in order to be counted, shall be signed by a responsible party and accepted by us. This proposition is open to and includes all business contracted in accordance herewith and the printed instructions to be furnished on application at our offices up to and including July 1, 1911. Everybody is eligible to accept this offer except employes of the Subway Telephone Construction Company and the Illinois Tunnel Company.

GENERAL CONDITIONS: Certificate of authority to solicit, subscription blanks, general information concerning service and other details, will be furnished on application at the Contest Bureau.

Apply immediately at S. W. corner Van Buren and Market Sts. Automatic Telephone No. 52-131.

Subway Telephone Construction Company

Bruce G. Moffat Colleciton

This May 2, 1911, advertisement in the Chicago dailies announced that long distance service would soon be available to automatic telephone subscribers.

Bruce G. Moffat Collection

Acting on behalf of the Illinois Tunnel Company, the affiliated Subway Telephone Construction Company conducted an aggressive advertising campaign directed at local businesses during 1911.

One Operator and One Hundred Subscribers

One operator, under the *old style manual method* of telephoning, must answer the calls of one hundred busy, hurried men. Is there any wonder that telephone service is demoralized in Chicago?

If your operator is attending to the wants of one of the other 99 subscribers, you must *wait your turn.* You must tolerate confusion, errors, wrong numbers, cut-offs, and congestion.

Automatic Telephone Service

is the twentieth century method—speedy, sure and secret. Your connection is made instantaneously, or if your party is talking you are notified as promptly.

The Automatic Service saves one business day a month. 40,000 in Chicago have subscribed for it.

Use your Automatic Telephone and **advertise** your Automatic number. **It will** mean more business for you. And use this **strictly private** service in your residence, too.

Telephone Contract Department, Automatic 32-525, and our representative will call.

SUBWAY TELEPHONE CONSTRUCTION COMPANY

164-166 Washington Street, CHICAGO

Bruce G. Moffat Collection

Larry Best Collection

Exchange 71 was housed in this attractive building at 625 E. 39th Street.

industry publication gushed: "To the Chicago business man it means the opportunity to test out the value of competition in producing a greater development and a better service at lower rates; and above all it means that he will be in touch with the 1,200,000 Independent telephone users who are in the trade territory naturally tributary to the Western Metropolis."

The annual charge for unlimited telephone service was $84 for business telephones and $50 for residential customers. The company soon realized that in order to secure enough subscribers its telephone lines would have to reach areas located well beyond the reach of the tunnels. As a result, the company announced that although its first exchanges were intended primarily for business customers, the initial telephone installations would be made in a fairly expansive area roughly bounded by North Avenue, Lake Michigan, 22nd Street and Ashland Avenue as well as in the Stock Yards district. To reach these areas, the company resorted to the use of normal-sized conduits and wires strung from poles. It soon turned out that the service area would have to be made even larger. By 1911 the system's southern extremity was Hyde Park (about 59th Street), while on the west side, a narrow strip extended to just beyond Kedzie Avenue.

Apparently, the solicitors were paid based on the number of contracts they returned to the company. Between April 1, 1910 and May 1, 1911, approximately 45,000 signed contracts were turned in by the solicitors. It was soon discovered that, in their zeal to recruit subscribers, the solicitors had misrepresented the facts, resulting in many persons refusing to pay their bills and demanding that their telephones be removed. As a result, the number of subscribers hovered near the 20,000 minimum mandated by the city instead of 50,000 or more that the company's officials had expected.

On June 5, 1911, the City Council ordered its Committee on Gas, Oil and Electric Light to conduct an investigation to determine if the telephone system was serving the required 20,000 subscribers. The city electrician reported that while the Illinois Tunnel Company had signed contracts for 44,040 telephones, its directory listed only 27,131 names of which 5,640 were duplications. Deducting these, plus 49 phones supplied to company officials, left 21,422 subscribers, technically fulfilling the franchise's requirement. Suspicions lingered as to whether all of those telephones were actually connected.

This report could, at best, be considered inconclusive because it was based solely on a review of the company's published directory and was not checked against other company records.

New Exchanges Opened

As the telephone installation work progressed, lines were strung to serve portions of the city's near north, south and west sides. Additional exchanges were opened while the now-technologically obsolete exchange on Fifth Avenue was closed. By July 1911, eight exchanges, having a combined capacity of 27,200 telephones, were in operation:

Exchange Name	Exchange Number	Street Address
Chronicle	31	175 W. Washington Street (space leased from newspaper)
Brooks	51	223 W. Jackson Blvd. (Brooks Building)
(?)	?	148 Maple St. (near north side area)
Cottage	71	625 E. 39th St. (near Cottage Grove Ave.)
Chemical	41	117 N. Dearborn St. (Chemical Bank Building)
Plymouth	61	327 S. Plymouth Ct.
West	81	Monroe & Paulina Sts. (northwest corner)
Yards	78	4170 S. Halsted St. (Chicago Union Stock Yards)

While most of these occupied leased quarters, Yards and Cottage were housed in buildings built for the purpose. With the expansion of the system, telephone numbers were revised to include numerical exchange prefixes (the exchange names were unknown to the public). Other prefixes were apparently added from time to time, but they were handled out of the existing exchange buildings. The company also installed a number of private branch exchanges (PBX's) for their largest customers including mail order giants Montgomery Ward and Sears Roebuck. PBX systems facilitated in-plant calling as well as allowing access to outside parties having dial telephones. Although data is lacking, the PBX installations probably had their own dedicated prefixes.

Among the factors limiting the growth (and profitability) of the telephone operation was the lack of connections with neighboring Bell-affiliated telephone companies. This lack of connectivity was touted as a virtue by stalwarts in the Independent movement who believed in unrestricted competition between telephone providers. At a practical level it meant that many Loop businesses had to maintain telephone subscriptions with both companies. For residential users it meant that you could not call your neighbor unless you both subscribed to the same company. Many smaller businesses and virtually all residential customers could not afford the luxury of two telephones and simply stayed with the Chicago Telephone Company.

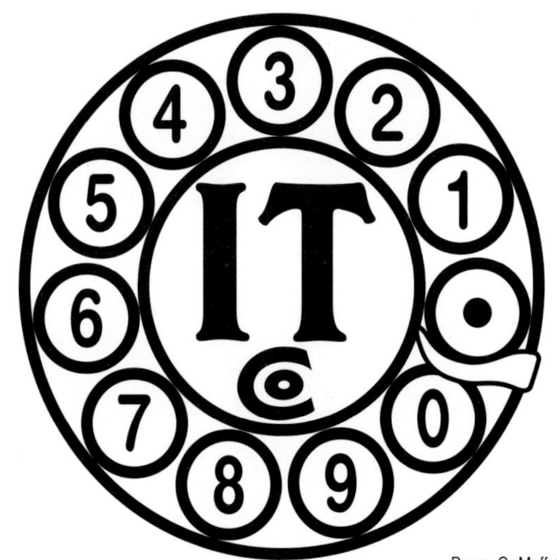

Bruce G. Moffat Collection

The only known company logo was this one that appeared in a few telephone-related advertisements.

Realizing that it also needed to offer long distance service if it were to remain competitive, the Illinois Tunnel Company's receivers entered into an operating agreement in early 1911 with the receivers of the Interstate Independent Telephone & Telegraph Company for long-distance connections between Chicago and various towns in Illinois and Indiana. According to the *Economist*, long-distance service was initiated on April 21, 1911, "to many of the more important cities and towns of Illinois." The relationship between the two struggling enterprises was made closer with the reported sale of large blocks of Interstate stocks and notes to a group of Tunnel Company investors. A merger of the two systems never occurred and little long distance traffic was ever realized due to the limited reach of the Interstate's lines compared to those of the Bell System.

The Telephone System Gets a New Name

In 1912, a comprehensive reorganization of the various companies was set into motion. The first step in removing the receivership was taken in early February when the U.S. District Court ordered the sale of the Illinois Tunnel Company's properties to take place in March. The fact that the company's combined telephone and railway operations were now profitable was borne out in comments made by president C.O. Frisbee and published in the February 24, 1912, issue of *Telephony*:

> It is contemplated that the reorganization will be effected and the property taken out of the hands of the receivers within the next sixty or ninety days. When the reorganization is finally effected the company will be upon a sound and

> conservative financial basis, with funds enough to carry out the purposes of the stockholders to extend the new automatic telephone system over the entire city of Chicago.

Frisbee, in a fit of misplaced optimism, also boasted that his telephone system would eventually have connections with every city in the country.

At about this time, the financial and corporate reorganization of the various tunnel-related properties began to take concrete form. The subsidiary Illinois Tunnel Company was replaced by the Chicago Tunnel Company on March 27, 1912. Chicago Tunnel then created its own subsidiary to manage the telephone system. Borrowing from the past, the name chosen for this short-lived subsidiary was the Illinois Telephone & Telegraph Company. The Chicago Utilities Company was organized to replace the Chicago Subway Company as the system's holding company on April 9, 1912.

Lingering Suspicions

Although management was busy projecting a positive image to the public, the city's aldermen continued to suspect that the company had not lived up to its franchise commitments. On March 12, 1912, the Committee on Gas, Oil and Electric Light sent to the full City Council a resolution declaring that the Illinois Tunnel Company's telephone franchise was technically forfeited because it did not have the required 20,000 stations installed by the June 1, 1911, deadline. Instead of taking immediate action, the aldermen referred the matter to a committee and authorized the expenditure of $5,000 to conduct a thorough investigation.

President Frisbee declared his willingness to cooperate in any investigation and then dashed off a pointed letter to the City Council protesting against the unfairness and injustice of the resolution. Frisbee pointed out that his company had never received a request to appear at any of the committee's meetings or to present its side of the case. A firm believer in competition, Mayor Carter H. Harrison struck a somewhat conciliatory tone by saying publicly that he would not approve any action that would result in the seizure of the company's equipment and turning it over to the rival Chicago Telephone Company.

Apparently unwilling to wait for the committee to complete its work, and mindful of the court-ordered sale of the company's assets scheduled for March 25, a group of aldermen presented a controversial resolution at the council's meeting of March 19. The resolution declared that the city had received information "proving" that the company did not, on June 2, 1911, have 20,000 bonafide subscribers. The resolution called for a full investigation of the allegations including taking "...such action as [the city] may deem advisable in respect to forfeiture of both the company's franchise and plant and equipment." The aldermen also directed the city's corporation counsel to have their resolution read at the sale.

A reorganization committee representing the holders of Chicago Subway and Illinois Tunnel securities purchased the properties for the bargain price of $5 million dollars. Looking ahead, Harris said "The sale of the Tunnel properties to the reorganization committee means the early completion of a comprehensive and modern telephone system for Chicago and the surrounding territory."

In July 1912, the company's relations with the city took a rare positive turn when the City Council's Committee on Gas, Oil and Electric Light approved a report prepared by the Everett Audit Company that found that the tunnel company did have the required 20,000 subscribers after all. The committee also received a report from the corporation counsel stating to the effect that he did not believe a court would sustain an action looking to the forfeiture of the company's franchise, telephone plant or equipment.

Now freed from the distraction of proving it was conforming to the franchise's requirements, the Chicago Tunnel Company, through its new Illinois Telephone & Telegraph subsidiary, was now able to concentrate on competing against the Chicago Telephone Company. Newspaper ads heralded the addition of major businesses to the "automatic" service while small booklets were mailed to potential customers touting the advantages of the new technology.

In a rather unusual example of commercial marketing, the company hired C. W. Winkler, who devised an unusual (and perhaps irritating) number calling contest. Winkler described his brainstorm in one industry publication this way:

> We sent out letters to our subscribers telling them that on a certain day we would give them theatre tickets providing they called a certain number which was designated the ticket number. In those letters we showed 25 automatic numbers.
>
> People would call and ask if it was the ticket number and those answering would say, "No, this is Marshall Field's," or, "This is "Hillman's." In that way the public became acquainted with the fact that the larger firms throughout the city had automatic service. When they got the ticket number they gave their name and address and two seats at one of the theatres were sent to them. We gave away 25 pairs of seats every week for a considerable period. The traffic in this way increased about a hundred thousand calls per day.
>
> It is a question whether or not you would call that legitimate traffic. Some say no, and some say yes, perhaps...

One wonders what the harried grocery clerk at Hillman's thought of getting all of those unsolicited phone calls.

6 THE ECONOMIST. Chicago, U. S. A.

THE AUTOMATIC TELEPHONE

means additional telephone service. Its secrecy and instantaneous connection will do much to make it popular.

No business is so private that it cannot be safely discussed over the Automatic.

We are in position to give prompt attention to the installation of our service. Have our representative call.

A card will bring him.

ILLINOIS TUNNEL COMPANY

164-166 Washington Street, [Telephone Dept.] CHICAGO

Bruce G. Moffat Collection

This advertisement appeared in the Economist, *a local financial publication, on January 7, 1911.*

Besides contests and the usual distribution of calendars and personal phone directories, the company launched a number of special services to generate calling traffic. One that debuted in 1912 allowed sports fans to obtain the latest baseball game scores. As an added inducement, the company purchased and gave away 250 baseballs to people who called the service on 10 consecutive days.

Advertising activity was largely confined to advertisements in the local business weekly, *The Economist,* with perhaps an occasional insertion in one or more of the regular daily newspapers. Small informational displays were also set up in downtown store windows and included photos of prominent people such as Mayor Carter Harrison using one of the company's phones. Visiting actors were also enlisted to pose for photos with the company's product. All of this was done with the objective of impressing on the public the benefits of modern automatic telephone service.

For Sale

In June 1913, there was renewed speculation that the Chicago Tunnel Company, despite all outward appearances, was indeed very interested in selling its Illinois Telephone & Telegraph subsidiary to the Chicago Telephone Company. The basis for this speculation was a request made to the City Council to release the company from a provision of its franchise "which would tend to make competition inoperative." This language effectively precluded any sale of the telephone properties to a competing local telephone company. Officials of both telephone companies denied that any such move was contemplated. A Tunnel Company representative said the change was requested because that particular section of the franchise was hindering sale of the company's securities.

Some industry observers speculated that the company's request would soon be followed by one to sell the automatic system to its competitor. Their suspicions were partially borne out the following month when Chicago Telephone representatives appeared before a City Council committee to support the Tunnel Company's request. A *Telephony* editorial urged that the City Council "should compel the Illinois Telephone & Telegraph Co. to carry out the provisions of its franchise, and save the people of Chicago from a monopoly with its attendant exorbitant rates and poor service."

On July 14, IT&T attorney Daniel J. Schuyler appeared before the Committee on Gas, Oil and Electric Light to request permission to go ahead with the sale. This request came as a complete surprise to the aldermen. Upon hearing Mr. Schuyler's proposal, the committee's members plunged into a lively discussion as to the fate of the IT&T's subscribers. When one alderman asked if this proposal would mean the end of automatic service in Chicago, Schuyler handed the question off to a Chicago Telephone attorney who happened to be present. His response was that they might wish to purchase the equipment but not the telephone rights. Chicago Telephone had no interest in the automatic equipment, but saw this as an opportunity to acquire a competitor and possibly gain the right to string cables in the tunnels.

The aldermen took no action until it could be determined whether there were anti-trust implications and to also give the city electrician time to place a value on the telephone plant. Subsequent meetings were devoted to exploring alternatives for maintaining automatic service including linking the automatic and manual exchanges so that subscribers could use just one telephone to contact any party. The political maneuvering reached a new level in July 1915, when city attorney Stephen A. Foster told the aldermen that former corporation counsel Sexton's opinion was incorrect and that the city did indeed have a legal right to forfeit the franchise governing the automatic telephone system. The mayor's own special counsel, Walter L. Fisher, who was no stranger to the controversy, weighed-in with a conflicting opinion. Lacking a consensus, the aldermen solicited additional studies and opinions about what to do.

In the meantime the struggle to retain subscribers continued. Even though it was quite evident that Chicagoans were not interested in "the automatic,", as it was known, the company's representatives continued to go to great lengths to retain customers.

The aldermen then turned to engineering consultant Kempster B. Miller who undertook a relatively detailed examination of the system and submitted his findings to the aldermen in a September 30, 1915, report to the Committee on Gas, Oil and Electric Light.

Concerning the company's efforts to maintain its customer base, Miller wrote:

> If the charge for main line telephones appeared too heavy, [subscribers] were induced to change to party line service. If this in turn appeared too burdensome, the subscriber might change to a guarantee pay station, and if the "guarantee" feature proved objectionable, a non-guarantee pay station was forthcoming. In short, if the subscriber had no other fault than his propensity for non-payment, he was as welcome as a connected subscriber, if not as a paying subscriber...
>
> Whether the Company has ever had twenty thousand bona fide subscribers depends in part on the accuracy of the Everett Audit Company's investigation, and on the construction that the courts will give to the wording of the franchise, particularly as to the definition of "bona fide subscriber." This phase of the matter seems to be relatively unimportant now, since it is my understanding that the officials of the Chicago Tunnel Company have admitted their failure to retain twenty thousand bona fide subscribers. My own investigation, carried on during the past month, has shown me conclusively that the Company has now less than twenty thousand bona fide subscribers, regardless of what interpretation may be put upon that term. As already stated, however, the system is amply capable of serving twenty thousand bona fide subscribers if the Company could secure and retain them.

Miller's report also included a detailed analysis of the company's expenses, revenues and billing practices, as well as identifying subscriber calling patterns and the reasons given by former users for abandoning the service. Miller then offered his explanation for the system's failure:

> ... it must be concluded that from the financial standpoint the automatic enterprise in Chicago up to the present time has been a flat failure...

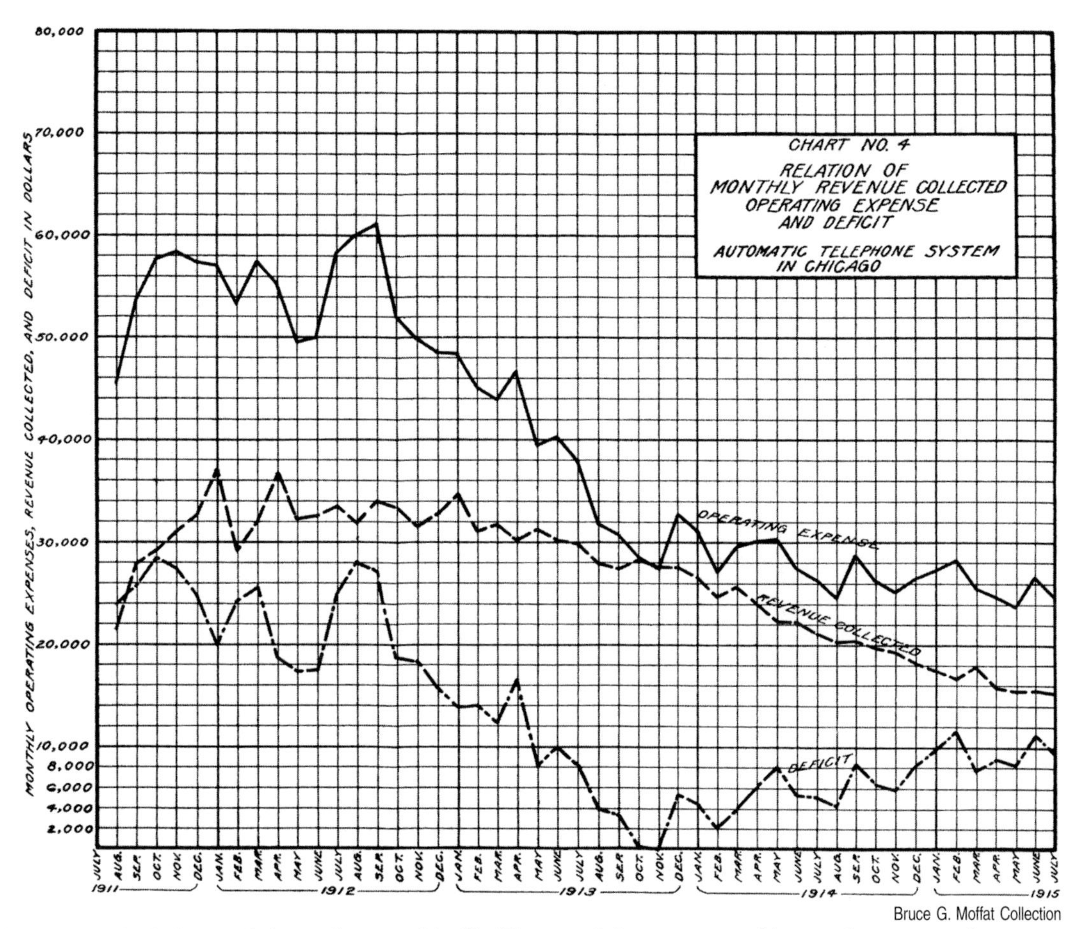

Bruce G. Moffat Collection

Miller's report included several charts that graphically illustrated the systems problems. Chart No. 4 (above) graphed the steady decline in reveneues (and a fluctuating deficit) for the four year period ending in July 1915. Chart No. 5 (below) showed the significant decline in traffic that had begun in mid-1912 had finally bottomed out at 1.8 million calls per month.

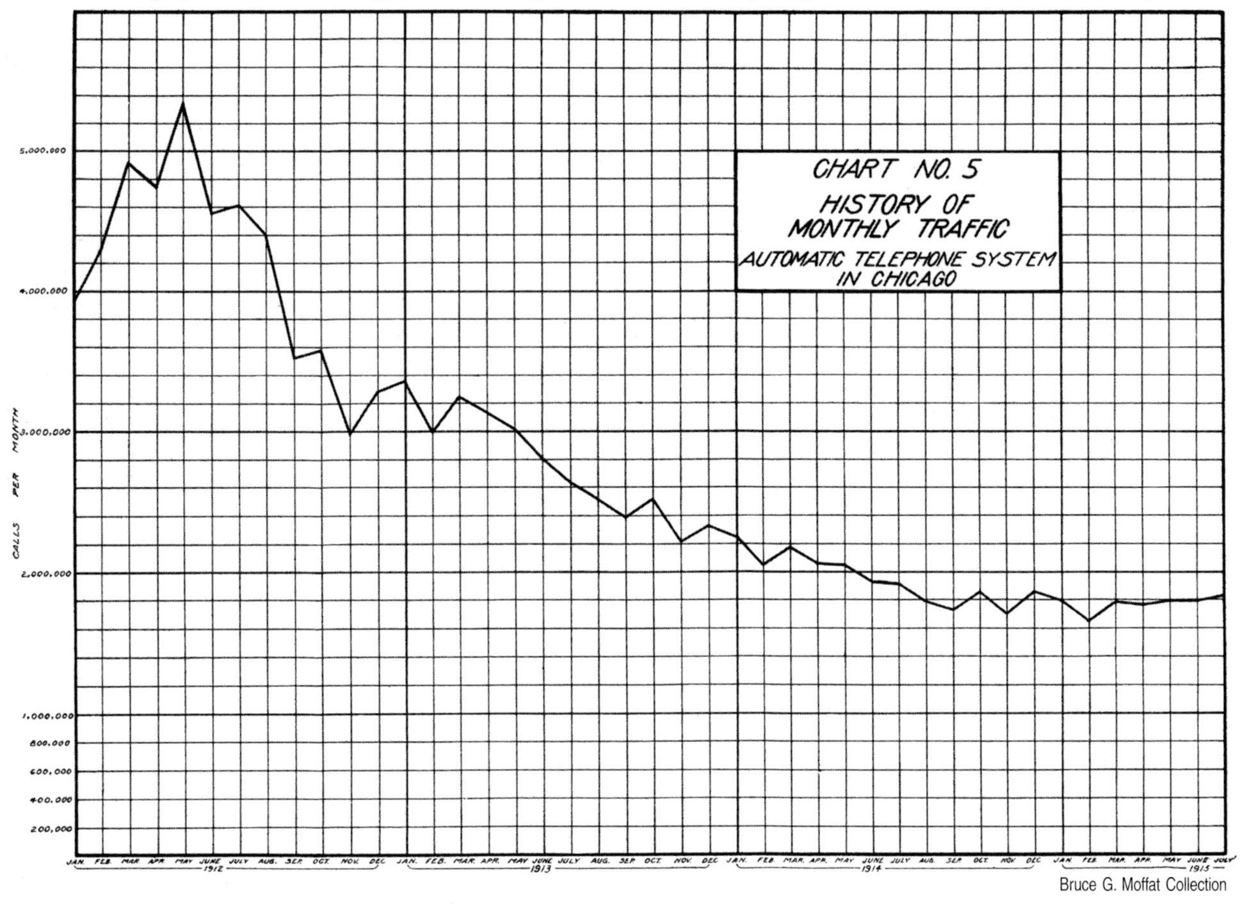

Bruce G. Moffat Collection

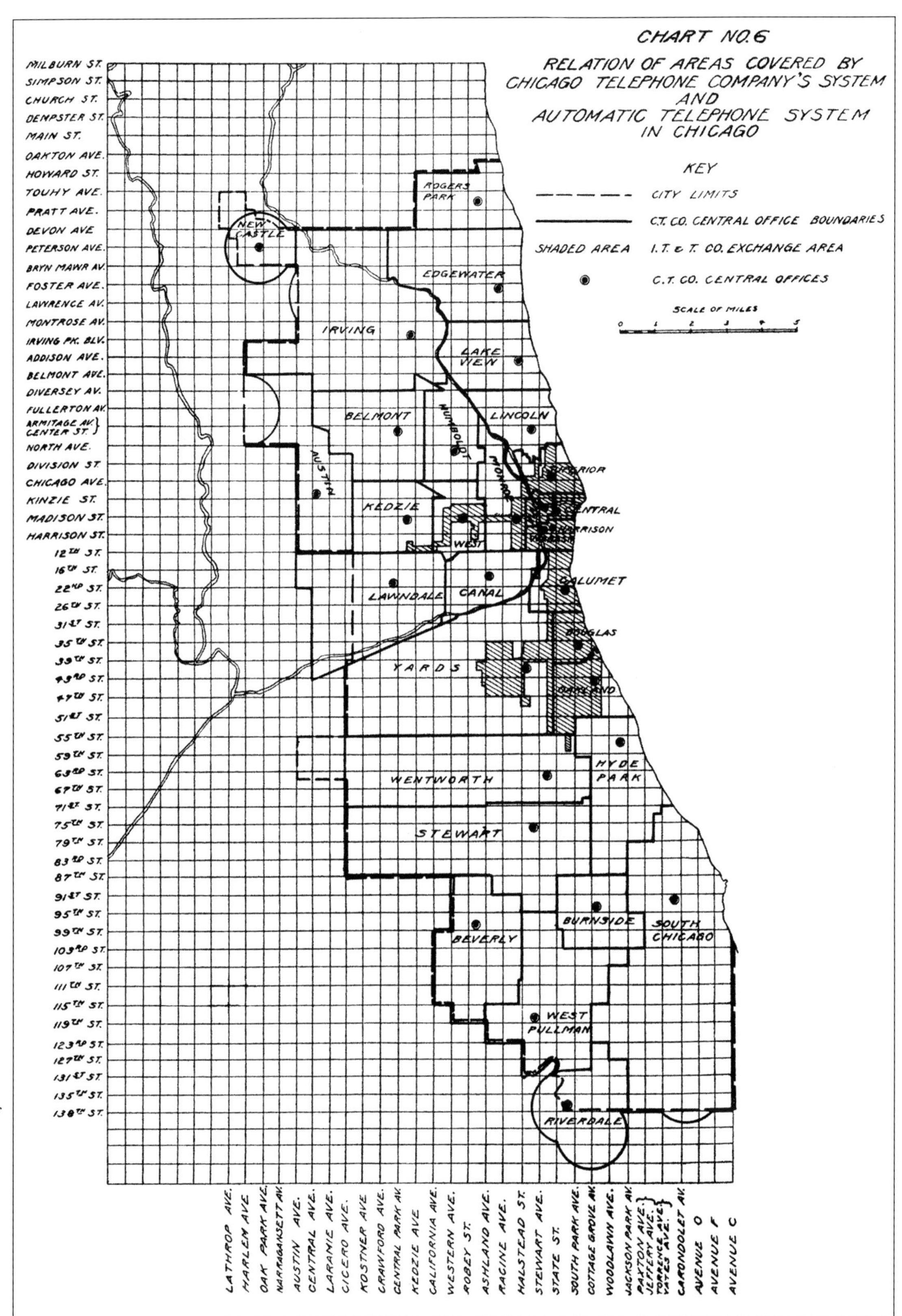

Bruce G. Moffat Collection

This map, included in Miller's report, shows just how limited the Illinois Tunnel/Illinois Telephone & Telegraph's service area was in comparison with that of the Chicago Telephone Company.

The automatic system, while of more recent development than the manual, has a history of about twelve years of successful operation in a number of cities in the United States and abroad. In such cities as Los Angeles, California, Columbus and Dayton, Ohio, and Grand Rapids, Michigan, the automatic system has proven its ability to compete successfully with the manual. I have had occasion in past years to make investigations in several cities of the relative popularity of the two systems... I have found in general, that in such cities... the patrons of both systems prefer the automatic. The results secured by other investigators have warranted the same conclusion.

It is obvious, therefore that the cause for the failure of the automatic in Chicago cannot be attributed to the inability of automatic systems to please the public.

If public acceptance of the technology was not the cause, then what was? Miller concluded that the Illinois Telephone & Telegraph Company's belated entry into the Chicago market was the reason:

> I believe that the fundamental cause for the failure of the automatic system here is the fact that the present automatic enterprise started at a time when the field was fully covered by an aggressive company, which had already covered the entire area of the city and had attained a high ratio of telephones to population.
>
> The disparity in the areas covered by the service of the two companies is clearly shown on Chart 6 [at left]...the disparity in the relative areas of the two systems was even greater in the early stages of the automatic development, at which time the automatic service did not reach much beyond the loop district.

Miller saw no hope for the enterprise and advised the city that any attempt to keep the system in operation would not be conducive to better or cheaper telephone service. Miller recommended that the company be allowed to sell the operation to the American Telephone & Telegraph Company (owner of the Chicago Telephone Company), which presumably would phase it out of operation as quickly as possible. (In 1913, AT&T had offered to pay $6.3 million for the struggling telephone system but was unable to secure the needed aldermanic support.)

Pulling the Plug

By January 1916, the two telephone companies had drafted a revised ordinance to allow the sale of the automatic plant. Following another round of acrimonious debate, the council approved the sale at its March 11 meeting by a vote of 46 to 22. Under the ordinance, Chicago Telephone agreed to pay $6.3 million for the telephone plant, which had been valued at just $2.2 million by the city. In addition, the Chicago Tunnel Company agreed to pay the city $500,000 for approving the sale. The state gave its approval on June 6, 1916, but not before extensive hearings were conducted to address the concerns of numerous independent phone companies about continued access to the lucrative Chicago market.

The final step was to obtain the approval of the U.S. Attorney General. The nation's chief law enforcement officer refused to give an opinion as to whether the purchase would violate anti-trust laws. When a last-ditch meeting on September 20 failed to clarify the areas of concern, the parties called off the deal. By now the system was incurring hefty losses, and the anticipated benefits to both parties from the transaction were diminishing almost daily.

Three months later, the City Council attempted to revoke the Illinois Telephone & Telegraph's franchise. A compromise was soon worked out, however, under which the company paid the city $200,000 in return for permission to abandon service. Finally, on July 28, 1917, a short article in *Telephony* confirmed the obvious:

> **Chicago Automatic Telephone to Quit July 31**
>
> Fifteen thousand automatic telephones in Chicago will cease operation and 150 employees will be thrown out of work at midnight on July 31, when the Illinois Telephone & Telegraph Co., a subsidiary of the Chicago Tunnel Co., will pass out of existence, in accordance with an order entered by the city council on December 15, 1916, forfeiting the franchise of the company....
>
> We figure it will take us about six months to take out the 15,000 telephones," said J.C. Payton, secretary and treasurer. "Our equipment is to be sold as rapidly as possible."

Telephone service was terminated on schedule, however last minute legal action by the Illinois Attorney General delayed the planned dismantling until September 13th, when a judge affirmed the city's right to terminate the company's franchise. The days of competing local telephone systems in Chicago were finally at an end.

As a footnote, the Illinois Telephone Construction Company, which had played an active role in the construction of the tunnels and in the building of the early telephone plant, continued to file annual reports with the state through 1918, although it had long since become a dormant corporate entity. In 1919, the state canceled the company's charter for failure to file an annual report and the following year sent it a bill for $309.75 in delinquent franchise taxes. Unable to collect, the state petitioned the Superior Court of Cook County, which entered a decree dissolving the dormant corporation on October 25, 1926.

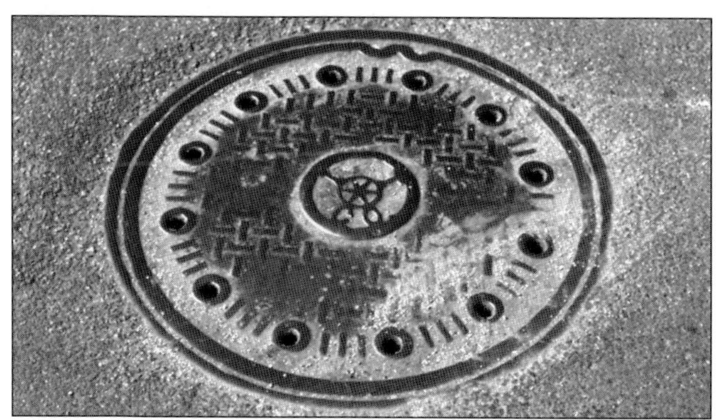

Bruce G. Moffat

In 1983, more than six decades after the telephone system was abandoned, this Illinois Tunnel Company manhole cover was still in place at the corner of Wells and Illinois.

5 Freight Operations

Commodities handled through the freight tunnels can be loosely grouped into just four categories: less-than-car-load (LCL) freight which was normally shipped in wooden crates or barrels; mail; coal; and ash/spoil. Although coal and ash/spoil traffic could at times be considerable, the company relied on LCL traffic for the bulk of its revenues. Mail proved to be a short-lived and costly commodity and was given up after just a few years.

LCL Freight

Major LCL customers included many of the Loop's downtown department stores. Major retailers such as the Boston Store and Marshall Field & Co. relied on the tunnel trains to not only deliver inbound cargo from railroad freight terminals and boat docks on the Chicago River, but to also handle outgoing mail order and home delivery shipments. Other customers included some area factories and warehouses that catered to the wholesale trade.

The company's promoters realized that if they were to more effectively compete with the teamsters for this traffic it would be necessary to establish specialized freight transfer and storage facilities. Such facilities would enable the successful solicitation of traffic from area businesses that were not connected to the tunnel system for one reason or another. Ideally, these installations would include space for loading horse drawn wagons (and later motor trucks) as well as railroad sidings and perhaps a boat dock.

The earliest attempt to establish a freight transfer facility involved the purchase of the sprawling river front dock and warehouse facilities of the Chicago Dock Company located at Taylor Street and the south branch of the Chicago River. Purchased on July 20,

Public Station No. 1, along with the headquarters for both the Tunnel Company and George W. Jackson, Inc., were housed in these two buildings located on Jackson just east of Halsted. Construction of the Kennedy Expressway later obliterated this site.

Bruce G. Moffat Collection

ADVERTISEMENT. 979

LOOK UNDER THE LID!

Tunnel Your Goods and Save Money

Look under the lid of Chicago's Loop—"the busiest square mile in the world"—and see what is going on forty feet below. The business traffic of the city is not all on the surface.

Busy Beavers

A veritable army of busy beavers are at work down under the Loop's lid—and many of them work both day and night. In the form of steel freight cars they are busily plowing their way over the 60-mile network of tunnel tracks constituting the transportation facilities of the

Chicago Tunnel Company

Quietly, efficiently, rapidly and safely these tireless carriers haul 2,400 tons or more of freight daily from the freight terminals of the railroads to the city's warehouses and office buildings. They will carry for a customer a single package or a carload or a whole trainload. They carry as willingly a lady's hat as a ton of coal. And they help to beautify the city by being the silent arteries through which flows the debris of razed buildings, excavations and other unsightly impedimenta.

There are four Public Receiving stations, all conveniently reached from the Loop, and there are hundreds of connecting stations within the Loop itself.

This system for freight traffic keeps 6,000 or more trucks off the already congested Loop streets during every business day.

Save Time—Save Money

This method of transportation is a saving to the shipper, for the railroads literally "pay the freight." Save your money and your time. Tunnel your goods and pull down your overhead.

CHICAGO TUNNEL COMPANY
CHICAGO WAREHOUSE AND TERMINAL COMPANY
754 W. JACKSON BLVD.
Telephone Haymarket 6300.

Ed Anderson Collection

After the abandonment of the telephone system, advertising was rarely done. This undated piece apparently appeared in a business publication in the late 1920's.

1904, for the then-considerable sum of $2.5 million from the Albert Dickinson Company (a seed wholesaler and distributor), the property, which spanned several acres, was the only non-railroad-owned river frontage in the immediate area and boasted several grain elevators, warehouses and railroad sidings.

The company had big plans for the parcel including construction of a mammoth 12-story warehouse (5 below ground and 7 above) measuring 404 feet by 699 feet. The ground floor of the structure was to be used as a vast covered yard for standard gauge railroad cars carrying commodities to or from tunnel customers. The lowest level was to house the Tunnel Company's own freight yard. Elevators would be used to bring cars to the main floor for the transfer of goods while the other floors would presumably be used for contract storage.

Public Stations

For reasons that are now unclear, this plan was abandoned and the property sold in 1912 to the Union Trust Company. In its place the company concentrated on a less ambitious warehouse concept that proved to be very successful. Known variously as "Public Stations" and "Universal Stations," these modest-sized freight terminals gave small off-line customers ready access to the freight tunnel system. The importance of these stations was not lost on the company's sales staff as they worked to lure business to the tiny trains. Located on the fringes of the system, these stations were technically operated by the affiliated Chicago Warehouse & Terminal Company (later renamed the Chicago Tunnel Terminal Company).

Upon arrival at one of the stations, a customer's

Public Station No. 2 on Kingsbury in the late 1920's.

Sveriges Television

Long after tunnel operations had ended, Public Station No. 3 on North Water served (in a somewhat truncated form) as a warehouse for the Chicago Tribune. This view dates from 1980. Due to subsequent redevelopment this site is unrecognizable today.

Bruce G. Moffat

wagon would be unloaded by Chicago Warehouse & Terminal Company employees and the cargo sorted according to class and destination. The packages were then loaded into the Tunnel Company's cars for delivery to an on-line customer or to a railroad freight house for further handling.

Through bills of lading were prepared for customers by the Chicago Warehouse & Terminal, which filed tariffs on behalf of itself and the Chicago Tunnel Company with both the Illinois and Interstate Commerce Commissions. The CW&T also handled billing and collection chores for both companies.

By 1910, four stations were in operation:

Station #1 was located at 754 W. Jackson Boulevard. This facility was originally owned by George W. Jackson who leased it to the Tunnel Company. The offices of Jackson's own engineering company (George W. Jackson, Inc.) were located in an adjacent building at 748 W. Jackson. In the mid-1940s, declining freight traffic resulted in this station's closure. However the Tunnel Company's general offices and main repair shop, which were also in this building, remained until 1948. The building was then vacated and demolished to make way for construction of the new Northwest (later Kennedy) Expressway.

Station #2 was originally located in a warehouse along the north bank of the Chicago River at 314 N. Dearborn Street. The Steele-Wedeles Company, a wholesale grocer, owned this building. The building's fifth sub-basement (which was actually seven levels below street level due to the building's main floor being above ground level to match a bridge approach)

Bruce G. Moffat Collection

Public Station No. 4 was housed in the sprawling Soo Line freight house at Roosevelt and Canal.

Bruce G. Moffat Collection

Inside Public Station No. 4. Once loaded, the cars are lowered by elevator into the tunnel.

included four tracks for loading or unloading cars. Two elevators allowed cars to be raised to any floor for further handling if desired. The Steele-Wedeles building also had a truck dock, steam railroad (C&NW Ry.) siding and a boat dock.

In 1925, the public station activities were relocated to an existing single story warehouse at 566 N. Kingsbury Avenue. This was probably done to place the public station closer to the many factories in that area that lacked tunnel connections. The Kingsbury location had been owned by the Rutland Transit Company, a lighterage company that specialized in transferring cargo between ships anchored offshore and riverside warehouses. Rail operations at this station were discontinued when merchandise and less-than-carload (LCL) traffic was halted in September 1956. The building was then demolished to allow for the construction of the Ontario/Ohio feeder ramps to the Kennedy Expressway.

The former Steele-Wedeles building was closed in 1982 and was later demolished to make way for a planned commercial/office/residential development by a Canadian investment group. At the time of its closure, the building housed a small restaurant, various offices, a secretarial school and a men's discount clothing store. Other floors were used for storage. The tunnel level, however, had survived virtually intact down to the overhead wire suspended from the ceiling. The development did not materialize and a hotel was eventually built on the site.

Station #3 was located in a nondescript single-story building on the Chicago River at 421 E. North Water Street. This was the first facility to be closed without replacement, happening sometime in the 1930s. The building, however, survived through the 1980s and was used as a newsprint receiving station for the Chicago *Tribune*.

Station #4 was located in the Soo Line Railroad's massive freight house located at 507 W. Roosevelt Road. It remained in use until LCL traffic was discontinued in September 1956. The building was replaced in 1998 by a shopping center.

Monticello Realty Company

This 1930's view looking west was taken from the Chicago & North Western's freight yard at State Street.

The Steele-Wedeles Building

The Steele-Wedeles Company operated a wholesale grocery business out of a large warehouse building located on Dearborn at the Chicago River. Until the late 1920's, Public Station No. 2 occupied the lowest level of this building.

Bruce G. Moffat Collection

Looking north in the building's basement. Note one of the company's several inspection cars at left – apparently used to transport the company's photographer this day.

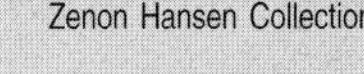
Zenon Hansen Collection

A 1924 cutaway view of the building showing that there were almost as many floor below street level as there were above. The lowest level had four tracks plus two elevators. The building also had boat and truck docks as well as a Chicago & NorthWestern siding.

Bruce G. Moffat

The basement as it looked in 1980. An abandoned flat car and other debris can be seen mired in about 6 inches of foul smelling muck. The flat is now at the Illinois Railway Museum.

Bruce G. Moffat

Looking northwest, the tunnel spur leads to the North Water tunnel. Ahead is one of the two elevator shafts that connected this level with the storage areas above.

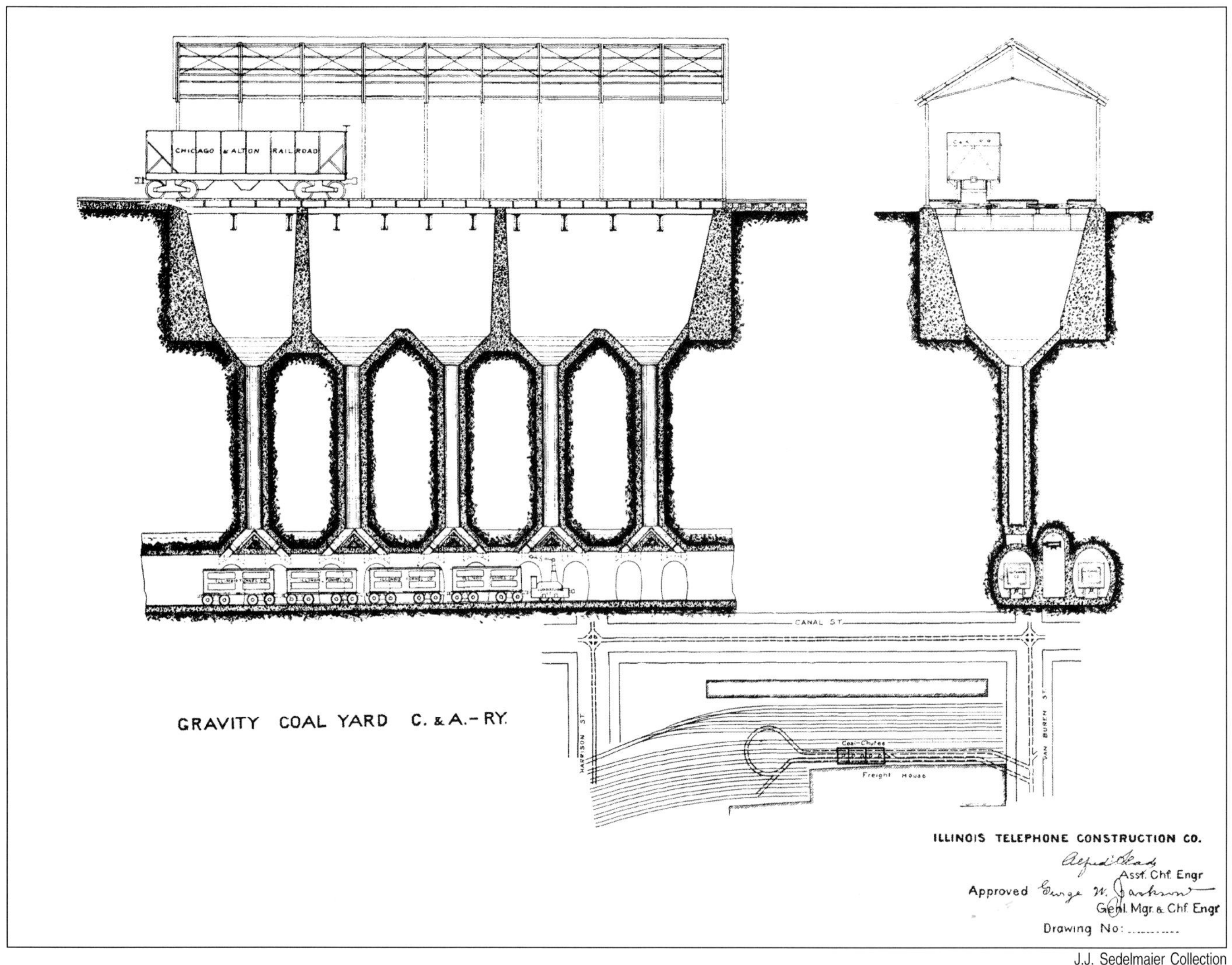

J.J. Sedelmaier Collection

Drawing showing the Chicago & Alton's coal transfer chutes near Canal and Harrison.

Coal

Coal traffic was a major source of revenue for many years. Construction was still going on in 1904 when the *Journal* published the following piece in its August 19 edition:

> Twenty-five hundred men are working night and day under the streets of Chicago on the tunnels of the Illinois (Tunnel) company, which officials now say will be completed and ready for operation not later than Dec. 1.
>
> At present the workmen are engaged in making connections between the old and new sections of the tunnel and President A. G. Wheeler said today he believed the work of delivering coal to downtown hotels and office buildings, which is to be the first commercial work of the company undertaken, will start Dec. 1.
>
> The electric railroad has been receiving dirt from new buildings in course of construction downtown, and rails, switches and crossovers are now almost all completed.

Wheeler's estimate proved to be overly optimistic. Freight operations, including the handling of coal, would not start in earnest until 1906, hampered largely by several contentious disputes with the city involving property tax assessments, the alleged building of tunnels and connections without proper permits, and settlement of buildings allegedly due to the tunneling.

Originally, the promoters envisioned construction of a large central coal receiving station on railroad property along the south branch of the Chicago River. Coal would be received from various railroads and reloaded into the company's fleet of special side-dump tunnel cars for delivery to on-line buildings that relied on the "black diamonds" for heating and hot water production. For some reason this plan was never implemented. Instead, four relatively modest coal transfer facilities were established.

Three of these installations utilized large underground chutes. The chutes were located near Van Buren and Canal Streets (Chicago & Alton Railroad), 14th and Dearborn Streets (Chicago & Eastern Illinois Railroad), and South Water Street east of Michigan

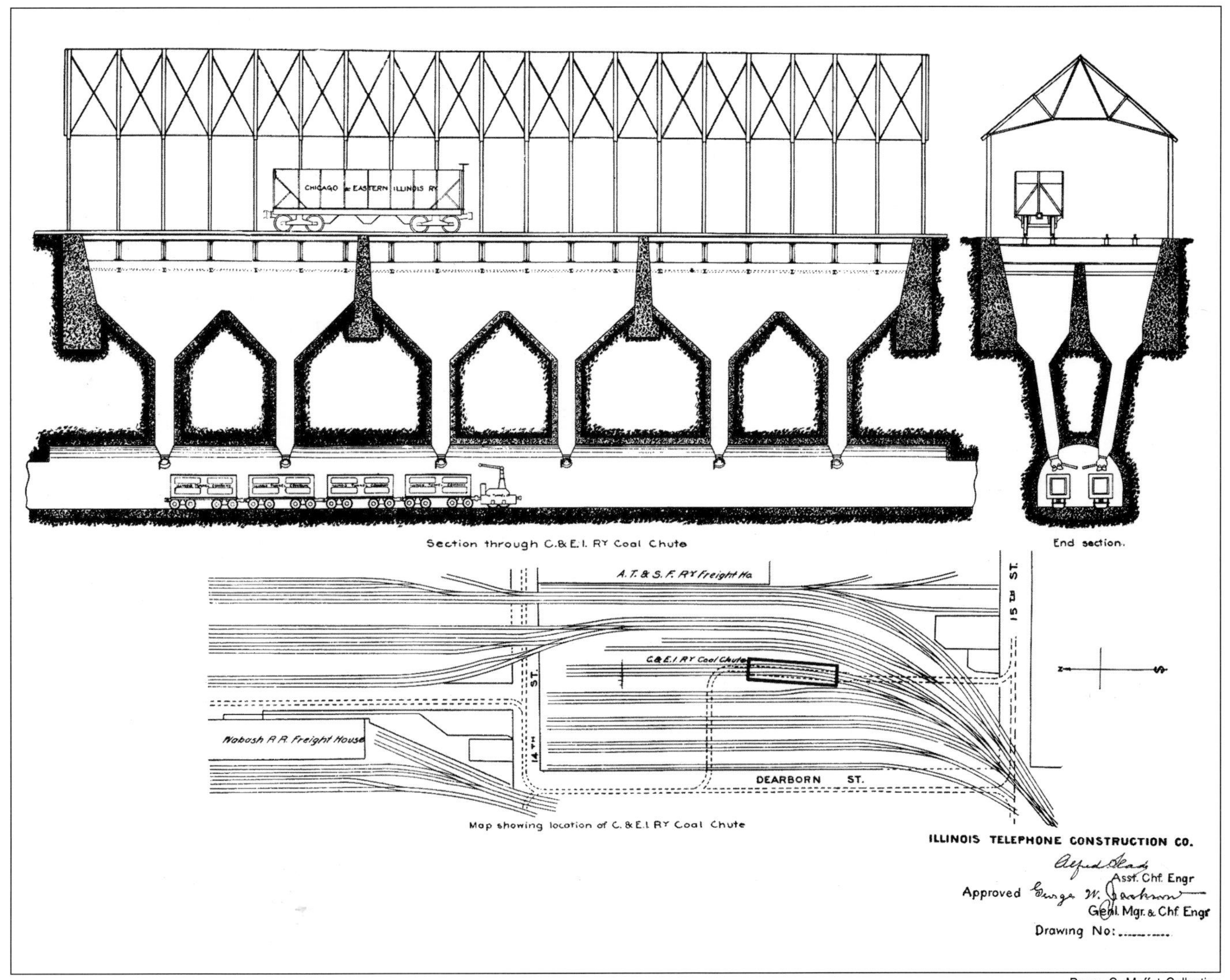

Bruce G. Moffat Collection

The Chicago & Eastern Illinois Railroad's coal transfer chutes were located just south of Dearborn Station.

Avenue (Illinois Central Railroad). This facility was actually owned and operated by the Crerar-Clinch Coal Company, the city's largest coal dealer.

The fourth station was located on surface trackage operated by the Chicago Warehouse & Terminal Company as an adjunct to its ash and spoil disposal station located at 13th Street and the Illinois Central tracks in Burnham Park. Apparently opened during the teens, this installation utilized a simple clamshell crane to transfer the coal. The loaded tunnel cars were then towed by one of the Warehouse & Terminal Company's Baldwin-built gasoline locomotives, or one of several modified ex-Tunnel Company electric locomotives, to an elevator shaft just west of the Field Museum and lowered into the tunnel. This operation was discontinued about 1927.

Upon arrival at the customer's building, the coal was unloaded onto conveyors which transported the fuel to the boiler room's bunker. This was hardly an economical procedure. Needless to say, the coal business was an early victim of motor truck competition. Trucks had the advantage of being able to simply dump their loads down a street level chute directly into the bunker. If this wasn't enough, some buildings chose to convert to natural gas, thereby avoiding not only transportation costs for coal delivery, but cinder removal and disposal as well. As a result, the Chicago & Alton chutes were closed by the early 1930s; the Crerar-Clinch chute closed in 1947.

By mid-1947, only one building, the Mandel Brothers Department Store, was receiving its coal by tunnel. Later that year, the Chicago & Eastern Illinois Railroad condemned the chutes on its property and notified the Tunnel Company that it would have to foot the bill for the necessary repairs if the chutes were to remain in use. Unable to justify the expense of keeping this facility open for just one customer, the CW&T discontinued the handling of coal on August 24, 1948.

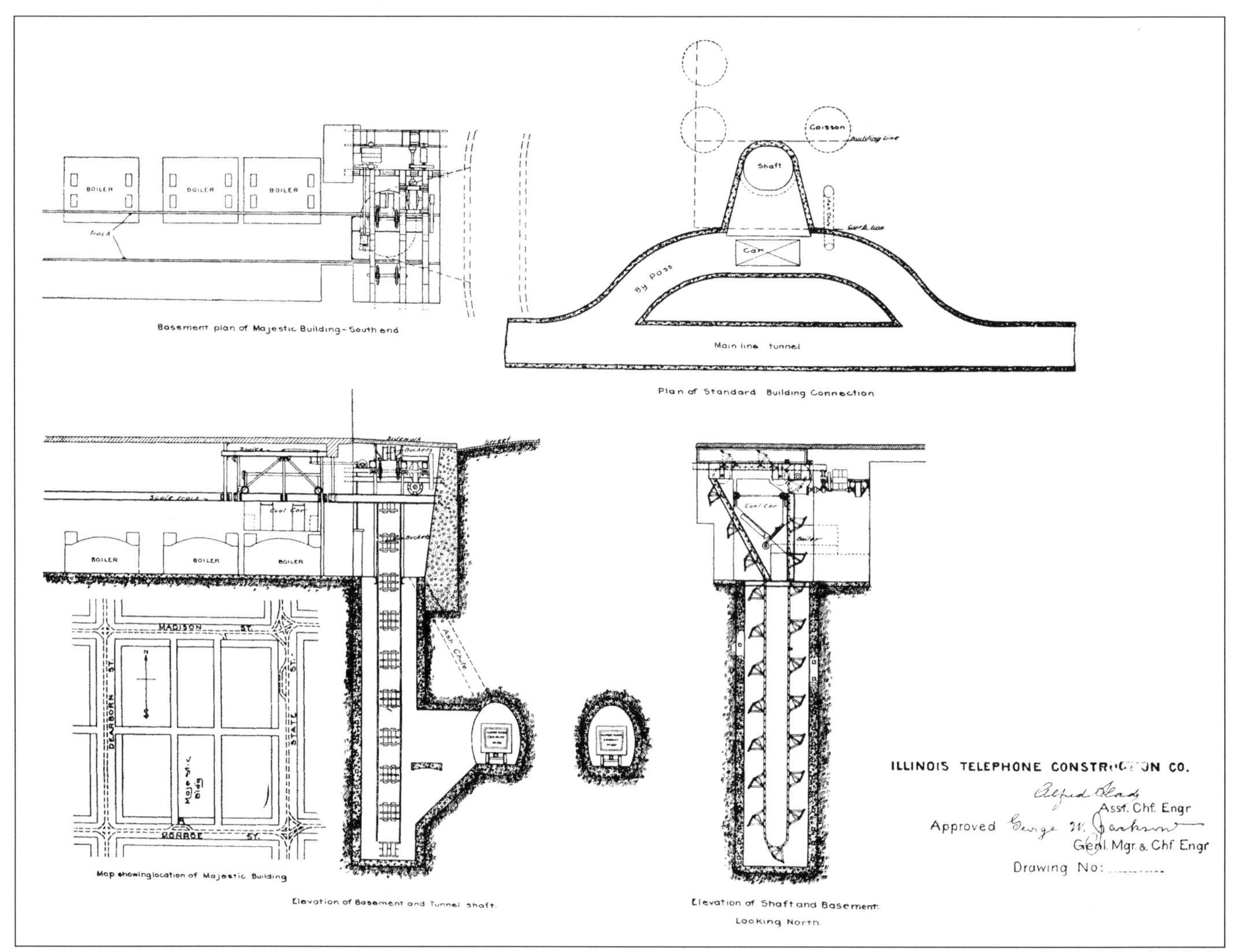

J.J. Sedelmaier Collection

At some buildings where the tunnel was located well below the engine room, a vertical coal conveyor had to be installed. The Majestic Building (home of today's Schubert Theater) was one such example.

The Majestic Building's tunnel interface was fairly typical of the smaller downtown buildings whose use of the tunnel railway was limited to coal and ash traffic. Incoming coal was dumped into the open hatch leading to the coal conveyor which hidden behind the recessed wall. The ash chute was located behind the photographer. Trolley wire has yet to be installed in this October 18, 1905, view.

Larry Best Collection

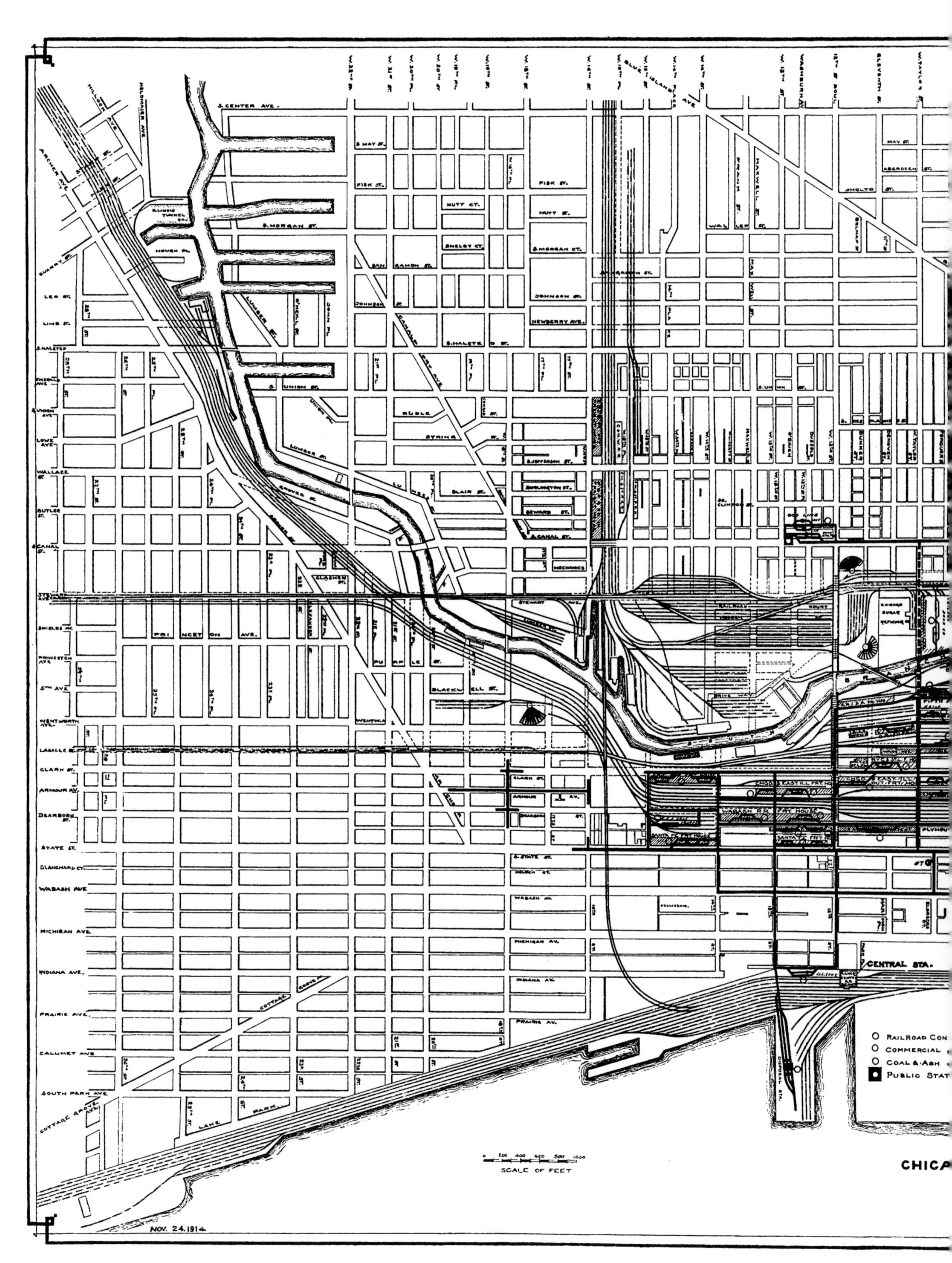

1914 system map. Note the inclusion of the isolated tunnel segments in the area of 16th and Clark Streets.

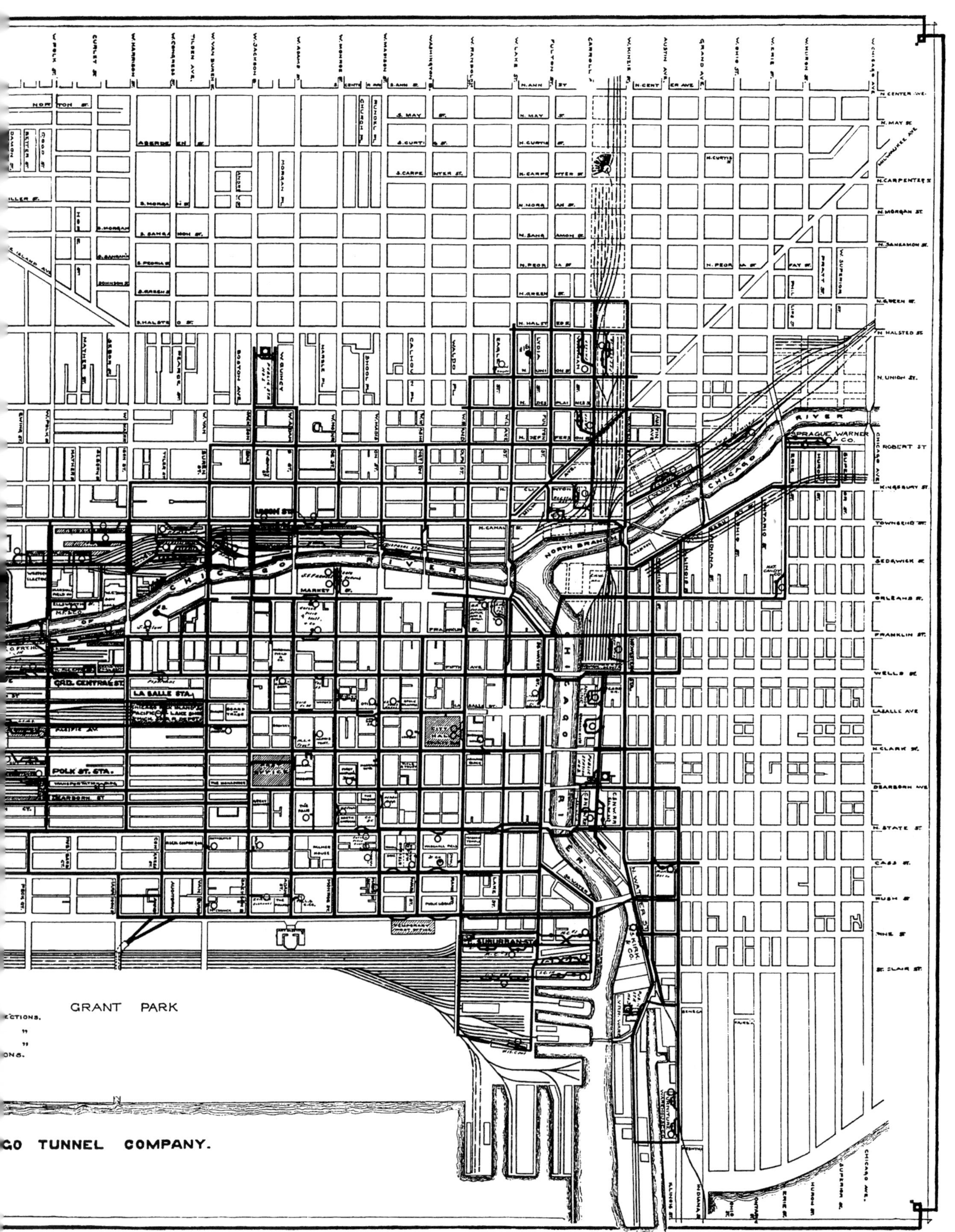

Bruce G. Moffat Collection

Ed Anderson

The spur leading to the Alton's coal chutes was marked by this simple sign.

Bruce G. Moffat Collection

A train of side-dump coal cars being loaded in the 1920's.

Bruce G. Moffat Collection

Coal cars being unloaded at a large office building. A conveyor was sometimes used to move the coal to the building's coal bunker for later use.

Ash

Even after the coal business had disappeared in the years following World War II, ash (cinder) removal traffic remained good. Since many tunnel connections with Loop buildings were below boiler room level it was more economical for these buildings to dispose of this commodity via tunnel train than to use trucks. Using gravity, the ashes were dropped from the furnace grates through chutes into waiting tunnel cars which were then hauled to a disposal station. Initially, ashes were disposed of in the same way as construction spoil – namely as fill at the Grant Park or Burnham Park landfills (until 1927) or transferred to a barge for dumping into Lake Michigan.

By the time that the Burnham Park site disposal station closed, the company had opened a new transfer facility on the west bank of the Chicago River just south of Grand Avenue. This facility was served by both barge and railroad and remained in service until the tunnels were abandoned in 1959. Coincidentally, ash was the only commodity being handled when the trains stopped running. Given the sporadic nature of freight movements in the final months, and the somewhat abrupt closure of the system, it is not surprising that decades later some of the final cinder loads were found languishing in the tunnels.

Ash car 526 and another unidentified unit are spotted under the recessed ash chutes of the Burlington Building at the corner of Jackson and Clinton in 1946. Note the detachable deflectors which directed the ash into the cars and not onto the floor.

Chicago Burlington & Quincy Railroad

View inside the second subbasement of Marshall Fields department store. The track passing through the doorway leads into a siding and the Wabash tunnel. Occupying the spur at right is an ash car. A telephone was conveniently nearby so the building's engineer could notify the Tunnel Company that the car was fully loaded.

Marshall Field's

Chicago Transit Authority Collection

Completed in 1906, Chicago's Federal Building was a massive yet attractive building. Covering an entire city block, the building's first two floors housed the city's main post office. This later day view looks northeast from the corner of Clark and Jackson.

Mail

In 1900, the Illinois Telephone & Telegraph Company had quietly broadened its corporate objectives to include the transportation of mail through the tunnels. Although it is unclear whether local postal officials or the company's promoters first conceived of using the tunnels for this purpose, both parties had, by 1903, determined that it was practical if not imperative. At that time, Chicago was not only a railroad center but also a center for mail going from one part of the country to another. Mail passing through Chicago would arrive aboard the cars of a given railroad at one of the city's six major passenger terminals where it was then unloaded. A veritable platoon of wagons was then used to transport the mail a mile or so to another railroad's terminal where it was reloaded.

Besides this traffic, mail originating or terminating in Chicago area had to be transported to the main post office, a temporary affair located in Grant Park at the foot of Washington Street. By 1904, the federal government was in the process of building a new federal office building in the city. A part of this massive building was to be used as the new main post office. But even before construction was completed it had become obvious that the space allocated for loading and unloading postal wagons was going to be inadequate. The tunnels promised to alleviate this problem.

One of the earliest references to the mail service appeared in the *Tribune* on March 3, 1904:

> Representative Boutell, who is deeply interested in improving the postal service in Chicago, made a strong argument before the committee on post offices...
>
> Mr. Boutell impressed upon the committee that the new post office facilities for wagon service are too limited and that something must be done to relieve the pressure, and that the problem confronting the post office department in transporting mails from the post office to the railway stations and from the railway stations to the post office may be solved using the tunnel of the Illinois Telephone & Telegraph company. He made a favorable impression upon the [congressional] committee.

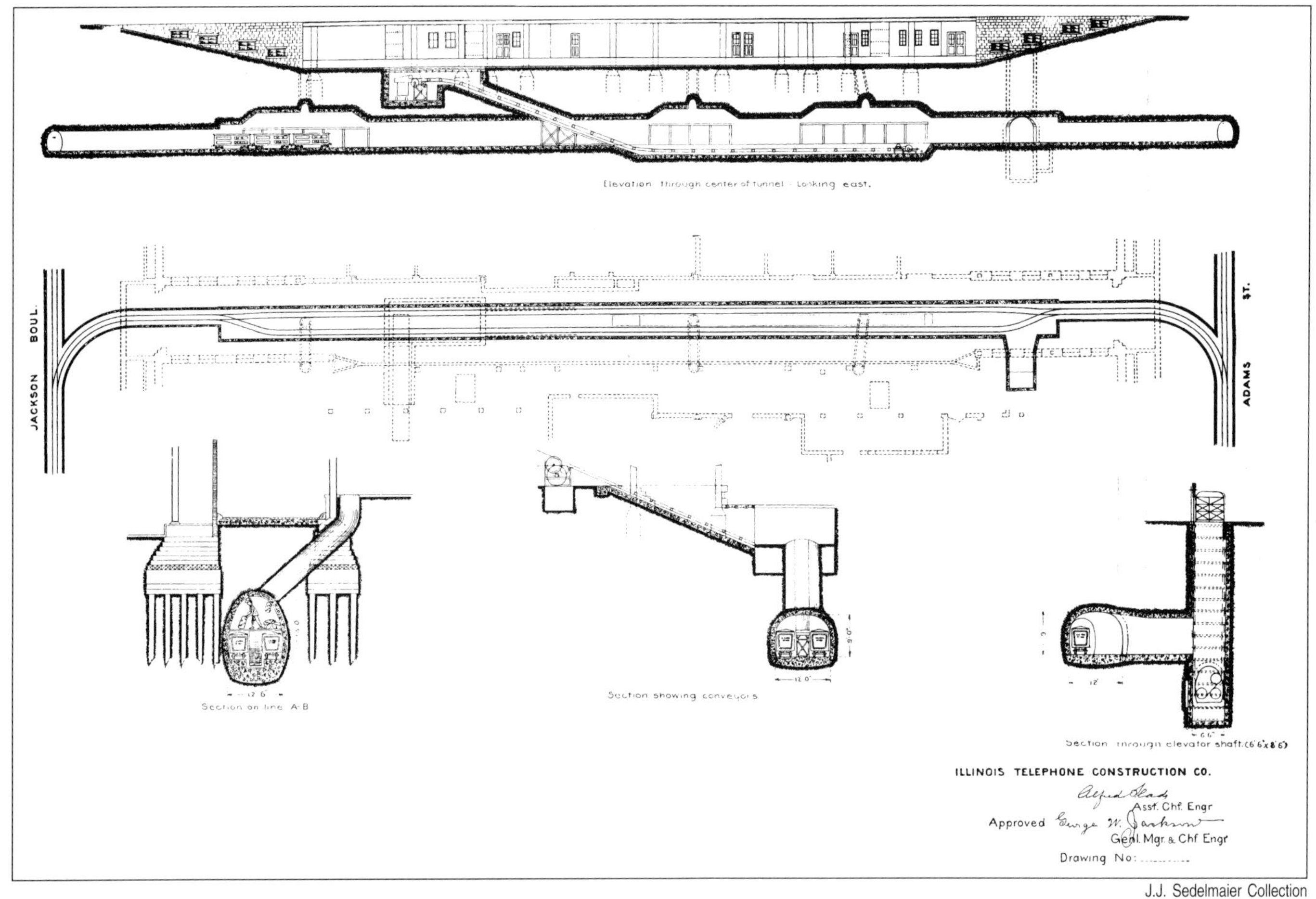

J.J. Sedelmaier Collection

A diagram showing the tunnel beneath the main post office.

J.J. Sedelmaier Collection

Postal officials and newspaper representatives gathered in the tunnel running beneath the city's new Federal Building to inspect the mail handling facilities on February 24, 1906.

Chicago Historical Society (ICHi-31230)

A conveyor was used to raise mail sacks into the building for sorting.

Another view beneath the post office. Note the large elevator at left.

Chicago Historical Society (ICHi-23293)

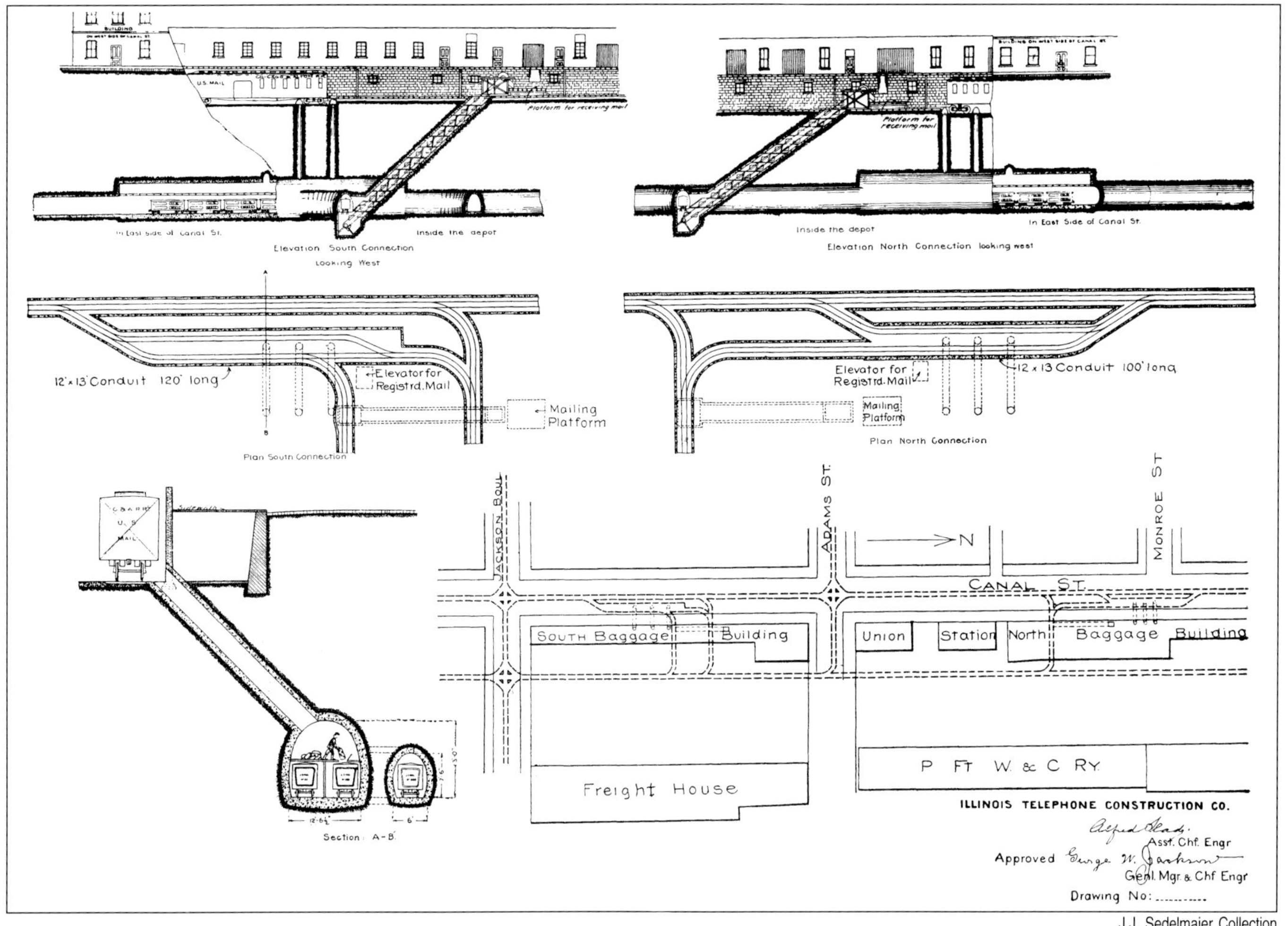

J.J. Sedelmaier Collection

A layout diagram of the extensive mail handling facilities at Union Station. After mail service was discontinued these tunnels were converted into a maintenance shop.

The U.S. Treasury secretary quickly announced his support and appointed a special commission to look into the matter. On February 16, the commission met with Wheeler and local postal officials and were treated to a tour of the tunnels and the various railway terminals. The result was an amendment to the still pending appropriation bill that stated: "Provided, the postmaster general, in his discretion, may contract for the performance of this service in Chicago by underground electric cars and wagons at a sum not exceeding the present cost of this service in that city."

Some questioned whether Wheeler's company could transport the mail for less than the wagons given that the teaming company was operating the service at a loss. Wheeler knew he could not operate the service for less. In fact, he proposed to bill the government $172,600 a year. According to the *Record-Herald,* this was significantly more than the approximately $106,000 the government was reportedly paying; however it was only $16,000 more than the actual cost of the wagon service. Wheeler argued that by giving his company the contract, the government would actually save money by not having to make an estimated $300,000 in alterations to the building to correct the capacity problems. Far from opposing a potential rival, the existing contractor welcomed the opportunity to lose the business. Instead of finding the transportation of mail to be profitable, the contractor had been forced into bankruptcy because the $107,000 value of the contract did not begin to meet expenses. As a result, a bonding company found itself operating the service until the contract's expiration.

But before any contract could be negotiated, or even a trial run made, the postal appropriation bill that was still before Congress would have to be amended. This turned out to be more difficult than expected. On March 22, 1904, the *Chronicle* reported the introduction of an amendment authorizing use of the tunnel "providing the cost of that service shall not be greater than the cost of efficient wagon service covering the same ground." Although this wording allowed considerable room for interpretation, the inclusion of the word "efficient" was apparently made to appease congressmen worried about paying more than was necessary. The authorized appropriation was increased to $200,000, half of which was to be used for leasing the tunnels and the other half for the service itself.

On June 26, 1904, Postmaster General Henry C. Payne visited the tunnels and came away favorably impressed. However, it wasn't until February 11, 1905,

In addition to conveyors and chutes, two elevators were used at Union Station to raise mail cars to special sidings located in sidewalk vaults next to the baggage rooms. Car 50 is being unloaded in one of the vaults.

Chicago Historical Society (ICHi-31228)

that the mail haulage contract was actually signed. The contract for $172,600 annually through June 30, 1908, called for the company to handle mail between the various railroad stations as well as the main and temporary post offices. Actual operation was delayed until 1906, because of difficulties in securing city permission to build some of the necessary connections.

So proud was the company of having won the postal contract that George W. Jackson hosted a special underground tour for a delegation of local postal officials on October 28, 1905. Photographs taken during that tour were later published in various trade journals and in company-issued promotional booklets. On February 24, 1906, the first test train was operated. Carrying a load of empty mail sacks, the train ran from LaSalle Street Station to the main post office. Officials announced their pleasure with the test; however, some contractual and procedural issues remained to be worked out. This delayed the start of service until mid-1906.

Later, a three-volume compendium on contemporary engineering, *Engineering Wonders of the World*, edited by Archibald Williams and published in London, took note of the unusual circumstances surrounding the advent of mail service with this somewhat fancifully related entry, which was apparently written around 1907:

> Two years ago the subway officials secured from the postal authorities a contract for the transportation of the mails. This contract was obtained only by convincing the powers that be against their will of the advantages offered by the new system. The main post office building in Chicago is the heaviest edifice in the city, and its enormous weight is supported on piles driven down to bedrock.
>
> When the engineer first approached the postal officials with a proposition to connect the building with the tunnel system, so that the mails might be carried from one office to another underground, they scouted the idea, saying that any such connection would undermine the building, cause a settlement, and probably result in its entire collapse.
>
> To prove that they were wrong, the engineer constructed a subway right under the center of the structure. In places this tunnel had to go through the piles on which the building rests, and the result is that the subway actually rests on the bottom portions of these piles and the top parts of the piles on the crown of the tunnel. After this work had been completed, the engineer invited the postal officials to take a trip through the tunnels as his guests.
>
> They accepted the invitation, and one can imagine their surprise when the car was suddenly stopped, and they were informed that they were right then under the post office building. For two days, it is said, the leading surveyors of Chicago were busy examining the edifice critically, but could not find a crack or a piece of plaster that had been disturbed. The post office still rests as solidly on its foundations as does the rock of Gibraltar. Not long afterwards the tunnel people secured the contract spoken of.
>
> For this mail work the railway employs 66 electric motors and 115 cars. In 1907 the electric trains made 337,060 trips with mails through the subways to the various branch offices, railway stations, etc. transporting 10,659,567 bags, pouches, and packages of postal matter. The record for this tremendous service was "99.51 per cent perfect"-that is to say, in this proportion of cases the mail was delivered at the proper stations in time. Last Christmas Eve the Company handled without a hitch, 44,341 bags of mails, 5,911 pouches, and 195 packages-a total of 50,447.

A Jeffery locomotive moves a mail car at LaSalle Street Station in 1906. Note the Strowger-type dial telephone on the wall.

Bruce G. Moffat Collection

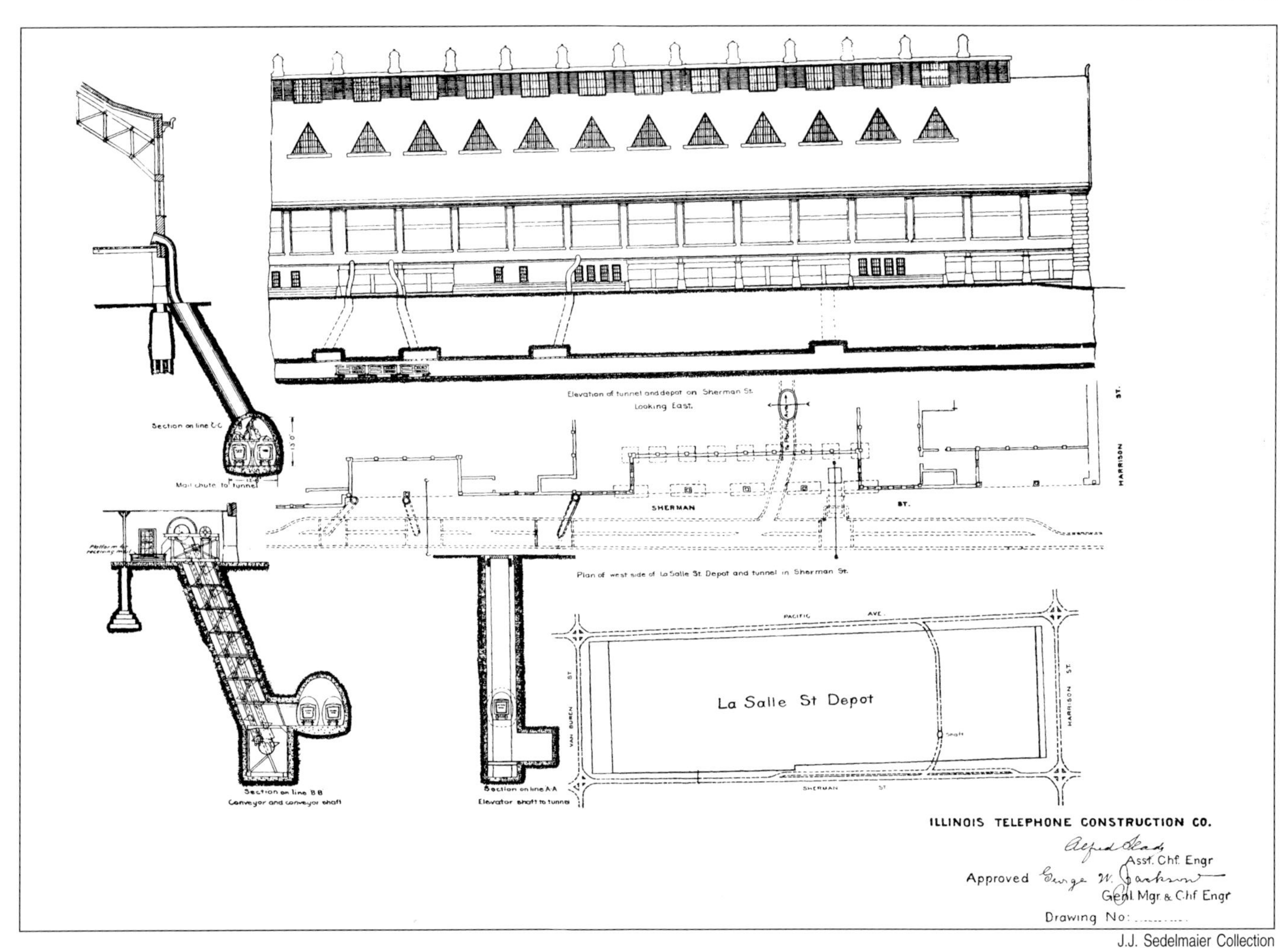

J.J. Sedelmaier Collection

The mail handling facilities at LaSalle Street Station included an elevator to raise tunnel cars as well as a conveyor and chutes.

Larry Best Collection

At Grand Central Station a Pere Marquete Railway post office car literally towers above tunnel mail car 59 in 1906.

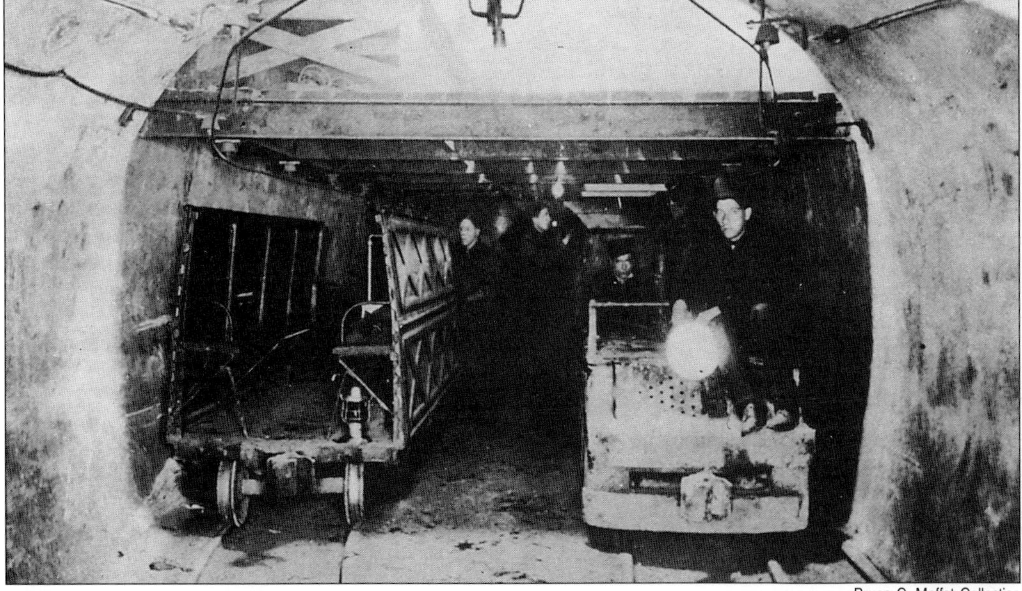

Bruce G. Moffat Collection

At Dearborn Station (Polk and Federal Streets) an elevator provided access to this upper level mail transfer area from which daylight could be observed.

The number of motors (locomotives) seems to be vastly overstated in view of the actual size of the fleet. Similarly it is questionable how many of the mail cars were actually built and used exclusively in this service. Registered mail was transported in specially locked cars with wire mesh sides while all other classes were handled in equipment similar in appearance to the company's standard closed-sided merchandise cars.

At LaSalle and Union Stations special underground mail handling facilities were established. At other railroad terminals the cars were simply raised to the surface for loading and unloading. Baggage movements between the railroad terminals were handled in a similar manner.

Mail operations were apparently discontinued with the contract's expiration in 1908. No doubt higher than expected operating costs arising from the need to operate numerous special trains and time lost waiting for late arriving main line trains or second sections were contributing factors.

Following the discontinuance of mail service these connections were closed. The facilities at Union Station were converted into a maintenance depot to augment the main shop on Jackson Boulevard. It would not be until 1951 that serious consideration would again be given to handling mail.

London's Post Office Railway

No sooner had the Illinois Tunnel Company exited the mail handling business than postal officials in London, England, began planning to build their own underground railway system. Then as now, London's narrow streets were extremely congested making the transfer of mail between railway stations and post offices both time consuming and expensive. In 1909, the Postmaster General appointed a committee to examine the feasibility of moving mail within central London by pneumatic tube or electric railway. Aware of Chicago's unique underground electric railway system, the committee visited the Windy City to take a look. Impressed, they returned to London and in 1911 issued a report recommending the construction of an underground electric railway.

Tunneling started in 1914, however World War I and a post-war depression forced a suspension of work until 1923. Train operations were inaugurated in December 5, 1927, using self-propelled cars built by English Electric. Additional rolling stock was added during the 1930's as mail traffic grew.

Like the freight tunnels, the Postal Railway was built to a gauge of two feet but differed in its use of a center third rail for power distribution and its reliance on driverless trains. Train speeds were regulated by varying the voltage (100 to 440 volts d.c.) supplied to the track from control panels located at the various postal facilities situated along the 23 miles of track. The system remains in service today.

Bruce G. Moffat

A 1993 view of the system's impressive maintenance complex located in the Mt. Pleasant post office.

Royal Mail

A train of 1927 English Electrics pauses at one of the postal station unloading facilities. Normally ran as three car trains, these cars were all retired by about 1980 in favor of more modern, higher capacity, equipment.

Bruce G. Moffat

One of these cars has been preserved by Royal Mail. Note the roll-out mail hamper.

General Train Operations

The motorman constituted the entire train crew. He was responsible not only for operation of the motor (locomotive), but also for loading and unloading coal, coupling and uncoupling cars, operating switches, and rerailing cars if necessary. Derailments were commonplace, especially in later years, as deferred maintenance began to take its toll on both track and rolling stock. Derailment sites were frequently marked by large gouge marks where wayward cars had come in contact with a tunnel wall.

From the start of operations, heavy reliance was placed on the use of the telephone for train dispatching. Some 266 telephones were installed at strategic locations throughout the system to allow motormen to communicate with the dispatcher who had to keep track of up to 300 train movements per day.

To compensate for the lack of passing sidings and to simplify operations, most tunnels were designated as "one-way streets." To identify the various streets under which the tunnels ran, brass street signs were installed at intersections. By the second decade of operation, the brass plates had been largely replaced with legends stenciled on the walls. Many of these markings are still legible today, bearing such long-gone street names as Austin Avenue (now Hubbard Street) and Market Street (today's Wacker Drive).

A 1911 article in *The Electric Railway Journal* provides some insight as to how trains were made up given that the system lacked the traditional freight yard:

> Each train is in charge of a single motorman, and assistants are placed at the originating and delivery [public] stations. Trains are made up at eleven assembly points by regular switching crews, each of which consists of two men.
>
> These assembly points are situated at street intersections with by-passes in the four quadrants. Their location is such that the switching crew has a certain number of shippers and freight houses from which to receive and deliver all cars. Trains are made up of eight and ten cars each, each assembly point being permanently located and under the protection of the signal system, and this makes switching on the main lines safe so far as approaching trains are concerned. When a local train approaches one of these assembly points it has either loads or empties to deliver to some shipper or freight house, depending on its direction.
>
> The entire system is divided into four districts, and all the cars to be delivered to any particular district are assembled in one train to be set out at the different delivery points indicated by the billing...

Considering the complexity of the system, it is interesting to note that only a minimum of signaling

Bruce G. Moffat Collection

A motorman uses one of the Strowger-type telephones to communicate with the dispatcher in 1911.

Ed Anderson

After the automatic telephone system was abandoned in 1918 dial-less instruments supplied by the Automatic Electric Company were used for train dispatching. This particular telephone was at Union and Carrol.

Right: *In 2000 at the corner of Clinton and Lake two generations of stenciled street legends dating from Tunnel Company days, along with a modern sign installed by the city, can be seen.*

was in place to protect trains. A July 28, 1905, article in the *Railroad Gazette* asserted that the system was to be run by "sight and sound." In 1912, the *Electric Railway Journal* reported that a fairly complex block signal system had been installed:

Bruce G. Moffat Collection

At some locations space occupied by an unneeded spur was converted into a reporting location where train operators could telephone the dispatcher for instructions and complete any required paper work.

FORM 55

SWITCHING REPORT

DATE 7-21-32

CAR NO.	FROM	TO	REMARKS
	4		
	–		
	–		
	–		
5445	–	14	
4434	–	7	
2292	–	73	
2544	–	73	
~~4654~~	–	~~16A~~	
4122	–	5	
5141	–	5	
3839	–		
4621	–		
6126	–	18	11
4680	–	2	(10)

Ed Anderson Collection

Motormen recorded their activities on a switching report card like this one. This report from July 21, 1932, indicates that the motorman picked up the listed cars from location 4 which was Public Station No. 4 and dropped them at various locations. All destinations were given numbers, unfortunately a listing by number and name has not survived.

Ed Anderson

Ed Anderson

The reflectors used to protect major intersections came in both green and yellow versions. Some were manufactured by the Nachod Signal Company.

> The system that was installed...was designed by S. Stolp, electrical engineer for the tunnel company, assisted by W. J. Kenyon, formerly vice-president. It consists essentially of a trolley contactor connected to light signal circuits so that front- and rear-end protection is obtained, as well as protection against trains approaching on an intersecting track. The signals and contactors are installed about 150 ft. from each junction point.
>
> The trolley contactor consists of a cast-iron box mounted upon the trolley wire and filled with an insulating medium such as Vaseline. A transverse shaft extends through the box, being journaled inside its side walls. The arms hang down from the ends of the shaft in the path of the trolley wheel or harp. This shaft carries a circuit-closing member inside the box, which operates a relay. When the extending arms are struck by the trolley wheel the circuit leading to the relay from the trolley wire is closed. The rail connection of the contactor also has its contact members working in oil and is so constructed as to be unaffected by water or impacts it receives from the passing car wheels...The contactor is located a short distance outside the block, and the signal just precedes it. When a train leaves the 300 ft. block it clears all signals through the off contactor on the rail.

This elaborate system apparently did not remain in operation for very long, as no further references are made to it; subsequent accounts mention only the "sight and sound" system.

River crossings were equipped with illuminated cautionary signs displaying warnings such as "STOP" and "DRIFT HAS RIGHT OF WAY." A motorman could easily observe these signs as his train approached an undercrossing. Some sidings and intersections were also equipped with warning lights and signs.

The most common form of train protection at intersections, at least in the later years, was provided by green (or occasionally yellow) reflectors mounted on the wall at the point of intersection. These reflectors were aimed so that the light from an approaching locomotive's headlight would be captured and reflected down the intersecting tunnel, thereby warning approaching trains to proceed with caution and stop if required.

In view of the simple signaling and dispatching system employed, it is not surprising that the company was extremely safety conscious. Internally illuminated warning signs were installed at locations where it was determined that extra caution was required. Although subject to great variation in message and colors, these signs are known to have used red, green and black as background colors painted on white glass with unpainted raised lettering. Typical messages included: "CURVE AHEAD," "GO SLOW," "STOP," and several variations of "LOOK OUT FOR CARS" which included the name of an intersecting street.

Bruce Moffat

Loooking north at Randolph and Clinton, the reflector shown at the right in this photo protected the intersection.

Bruce G. Moffat

This warning sign was located on Randolph near Franklin.

Bruce G. Moffat

In the 1940's, this cautionary sign reading "What's your hurry? Make the tunnel safe" was installed on Jefferson near Kinzie. The tunnel segment behind the sign had been abandoned due to subway construction.

Ed Anderson

Long after the trains stopped running, this well-worn sign proclaiming "Any job is a safe job for a careful workman" was still standing silent sentry at Clark and Roosevelt.

Bruce G. Moffat Collection

Freight trains had been operating only a few years when this train was photographed leaving Marshall Field's department store and entering the State Street tunnel.

Freight cars were frequently stored in normally dormant tunnels because of the absence of a dedicated car storage yard. Since none of the company's freight cars had brakes to hold them in place, extra caution was required when crossing a tunnel where they were parked. Locomotives were equipped with screw-type brakes that were regulated using a railroad-style brake wheel.

Also installed at various points on the system were a series of very large internally illuminated triangular signs. These signs bore a variety of catchy safety slogans such as: "WHAT'S YOUR HURRY? MAKE THE TUNNEL SAFE" and "BETTER BE CAREFUL THAN SORRY." Besides coming in a variety of colors, these signs used a mixture of type fonts.

While many buildings were connected to the

Bruce G. Moffat

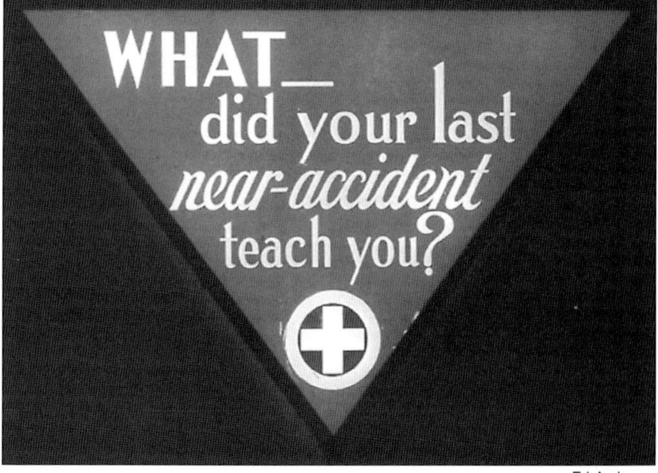

Ed Anderson

Messages that need no explanation.

Chicago Historical Society

During December 1929, a Chicago Daily News *photographer recorded this train being loaded with Christmas purchases at an unidentified department store (perhaps Mandel Brothers).*

tunnel system by elevators, others were built to depths sufficient to permit the operation of trains directly into their basements. In these cases, the overhead trolley wire was often extended into the buildings as well. In the case of the Marshall Field & Company's State Street store, which occupied a full city block, it was possible for a train to enter by way of State Street and exit onto Wabash. A few very large buildings such as the Field Museum and Merchandise Mart had locomotives assigned to them on a full-time basis to switch cars. In these cases, building personnel normally operated the locomotives, but then left the assembled string of cars for pickup by a regular train.

Extensive use was made of automatic elevators to raise cars into buildings whose lowest levels were nonetheless still above tunnel grade; this in an era when automatic elevators were not in common use. On several occasions in the early years, employees who were not completely familiar with the operation of automatic elevators sustained serious injuries while positioning cars.

To accommodate the servicing and maintenance of the fleet, which in 1929 contained 149 locomotives and 3,280 cars, two repair depots were kept busy. Major overhauls and heavy repairs were made in the company's main building at 754 W. Jackson. Running maintenance and light repairs were handled in the old mail facility under Canal Street at Union Station, which was dubbed the "roundhouse."

At the roundhouse, separate maintenance areas were provided for locomotives and cars. The locomotive shop was located in a large bore at Canal and Monroe and consisted of a two-track shop area complete with a maintenance pit, hoists and work benches. Freight cars were looked after in a second large bore situated between Adams and Jackson. The car shop likewise had two tracks but lacked hoists. Several small tunnels that had been used for routing mail trains under Union Station were blocked-off to create locker rooms and work areas. It is unclear if there was any direct access from Union Station following the conversion from a mail transfer area to a repair depot. Absent such access, tunnel employees would have had to gain access through another nearby building.

Surplus locomotives were stored in the Monroe Street tunnel west of Canal and in other nearby bores between assignments. Following the closure of the main shop in 1948, in anticipation of construction of the Northwest (Kennedy) Expressway, all repairs were handled at the Union Station site.

SECTION
III
The Chicago Tunnel Company

6 Deficit and Default

In an effort to end the railway's receivership as expeditiously as possible (and preserve their investment) the Armour-Harriman syndicate appointed three New York financial experts to act as a committee to develop a reorganization plan for the several tunnel-affiliated companies. It was quickly determined that some $6 million would be needed to satisfy the combined indebtedness of the three firms.

Presented to the stockholders on November 14, 1911, the plan called for the formation of a new holding company called the Chicago Utilities Company to take the place of the Chicago Subway Company. Chicago Utilities would in turn control two railway-related subsidiaries: the Chicago Tunnel Company (replacing the Illinois Tunnel Company) and the Chicago Warehouse & Terminal Company.

Following acceptance of the plan, the property and assets of the Illinois Tunnel Company were foreclosed and sold on March 26, 1912, to a special purchasing committee, which then transferred them to the new Chicago Tunnel Company after its incorporation the following day. The sale's proceeds were used to retire the outstanding receiver's certificates and pay legal expenses incurred during the receivership. The assets of the Chicago Subway Company, which consisted of title and interest in the stock and properties of the two primary underlying companies, along with the Illinois Telephone Construction Company, were sold to a special purchasing committee on April 3, 1912.

The final step in the reorganization process took place on April 9, 1912, when the Chicago Utilities Company was incorporated and formally assumed ownership of the Chicago Tunnel Company and the Chicago Warehouse & Terminal Company. It was at this point that the new holding company organized another subsidiary named the Illinois Telephone & Telegraph Company (the second firm to carry that name

Left: *On February 26, 1927, a newsreel crew was busy filming freight tunnel activity from aboard one of the inspection cars.*

Bruce G. Moffat Collection

Sporting an improvised motorman's cab, one of the electrics assigned to Chicago Warehouse & Terminal for use at the Burnham Park "dump" site passes the brand-new Soldier Field stadium on December 6, 1926.

Sveriges Television Collection

Bruce G. Moffat Collection

A trio of views showing the Baldwin gasoline engines working in Burnham Park. For a while the surface trackage extended as far south as 31st Street. The bottom view appeared in a Spanish language Baldwin catalog.

Fred Ash Collection

Bruce G. Moffat Collection

in the tunnel's history), to operate the deficit-ridden telephone property. The Illinois Telephone Construction Company ceased to function as an active company at about this time.

The principals in these complex buying and selling maneuvers had also been affiliated with the original companies, rendering these corporate changes more or less cosmetic. Proceeds from the sales were used to satisfy debts, with the companies emerging from receivership in mid-1912. As was normally the case in reorganizations, the hefty investments made by the Armour-Harriman syndicate and the other bondholders were protected, while the investments made by persons through the purchase of stock in the now defunct corporations were wiped out.

Additional Lakefront Filling

The creation of Burnham Park by the South Park Commissioners was already well underway in 1912 when the Tunnel Company offered to provide spoil and ash at no charge to be used as landfill material. The fill would be dumped into the lake in the area around 13th Street to aid in the creation of badly-needed park land. More importantly, it would give the company an inexpensive and convenient way to get rid of unwanted material. The South Park Commissioners quickly accepted the offer.

To reach the Burnham Park site, the Tunnel Company built a tunnel east from a connection with the existing bores at 13th Street and Indiana Avenue, passing beneath the Illinois Central's tracks and their main Chicago station (sometimes known as Park Row) in the process. At a point just west of the present Field Museum of Natural History a large shaft was sunk and two elevators installed. Loaded cars were brought to the surface and towed over temporary trackage to the dump site by a small fleet of Baldwin gasoline locomotives operated by the Chicago Warehouse & Terminal Company. Operation of this newest of tunnel disposal stations began in August 1913.

Later, but prior to 1925, the gasoline locomotives were replaced by several Tunnel Company electrics that were assigned to the CW&T. For operation in "the outside world," makeshift cabs were added to protect train operators from the elements as well as flying debris. Landfill operations were discontinued with the completion of the parklands around 1930. More than 100 acres of park extending south from 12th Street were created in this manner.

Although the fleet numbers of the locomotives used are unknown, CW&T reported the fleet size to the Illinois Commerce Commission for several years:

Year	Quantity	Type	Builder
1917-1918	3	Gasoline	Baldwin
1919	4	Gasoline	Baldwin
1925-1927	7	Electric	?
1930-1930	1	Electric	?

Field Museum

This 1919 construction view shows the just-completed "tunnel" connection to the Field Museum's boiler room before it was covered over with earth. Excavation (ash) cars – and the IC's Central Station – are in the background.

The Field Museum

The Field Museum of Natural History was only one of many buildings that received coal and dispatched ash via tunnel. However, the construction of its connection with the rest of the subterranean system was unlike any other.

By 1915, the collection of the Field Columbian Museum (as it was then known) was rapidly outgrowing the confines of the former Fine Arts Building located in Jackson Park on the city's south side. This building, a relic of the World's Columbian Exposition of 1893, had also developed serious structural problems.

Deciding that the needed repairs would be too costly, the museum's board decided to construct a new building at the north end of Burnham Park. Coal for heating the building, and removal of the resultant ash, was to be handled by the tunnel trains, thereby keeping trucks off of the new park's roadways. Construction of the building began on July 26, 1915.

In order to reach the museum's new site a short connecting "tunnel" was built at ground level from the elevator shaft used by the disposal trains at 13th Street and continued southeastward to the southwest corner of the new building where the furnaces were to be located. Following the construction of the tunnel, it and the surrounding area were covered with an additional layer of fill, leaving only the Illinois Central's tracks at the original grade.

To maintain a connection with the surface disposal station trackage once the grade had been raised, a spur and ramp were constructed from the new museum tunnel next to the elevator shaft. Service to the museum probably began a short time prior to the building's public opening in 1921, and continued uneventfully for many years.

By 1947, the economic benefits of receiving coal by tunnel were negligible. The museum accepted its last load of coal by rail on February 14, 1947, with truck delivery starting exactly one month later.

Ash removal continued to be handled by the diminutive trains until June 1957, when the Tunnel Company discontinued service to this area. This ac-

Field Museum

Much remained to be done to Burnham Park when the Field Museum held its grand opening in 1921. The tunnel connection was at the right (southwest) corner of the building.

At least five ash cars are in the process of being loaded at an unidentified Loop building in 1928.

Bruce G. Moffat Collection

tion required the museum to install ash-handling equipment to facilitate removal by truck. Since there were no other active freight customers in the vicinity, the company was able to reduce its costs by not having to operate a train for just one customer as well as avoiding some track and power distribution system maintenance expenses. Left behind in the museum's connecting tunnel were Baldwin-built locomotive #508 and loaded ash cars 592, 653, 714, 766 and 856.

In late 1957, vandals broke into the small building on the west side of Lake Shore Drive that housed the elevator's hoisting machinery and stripped it of all salvageable copper. Thus, had the company been of a mind to do so, any attempt to retrieve the train would have been made prohibitively expensive. When the theft took place, the elevator machinery was still energized at 660 volts DC (on circuits separate from the tunnel's 250 volt overhead system). The building itself was demolished in 1966, and the shaft capped and covered over with earth.

Bruce G. Moffat Collection

Maintenance superintendent H. J. Sims pilots locomotive 511 west onto Adams from Clark in the 1930's.

Default

The renewed efforts of the Chicago Utilities Company and its subsidiaries to operate a profitable system were met by recurring annual deficits. Unable to make ends meet, the Chicago Tunnel Company and the Chicago Warehouse & Terminal Company defaulted on bond interest payments due to the parent Chicago Utilities Company, forcing the latter into default in April 1915.

In 1918, the Chicago Tunnel Company also suspended tax and franchise compensation payments to the city. During the next five years the situation did not improve. A minority group of Chicago Utilities bondholders, dissatisfied with the company's failure to meet its obligations, believed that the tunnel properties were being intentionally mismanaged to lower the market value of their holdings. They began to press for the money owed them. When the Armour-Harriman syndicate got wind of their intentions preparations were quickly made to sell the company.

The sale took place on July 15, 1921, in New York. The lone bid was made on behalf of the Armour-Harriman syndicate, making them the sole owners. The final step in the takeover process came on June 15, 1923, with the incorporation of the Chicago Tunnel Terminal Corporation in Delaware to replace the Chicago Utilities Company which was then dissolved. The issue of the delinquent tax and compensation payments due the city was left unresolved and would remain so until the negotiation of a new franchise in 1932.

With the reorganization of the various tunnel-related companies complete, attention was turned to

Bruce G. Moffat Collection

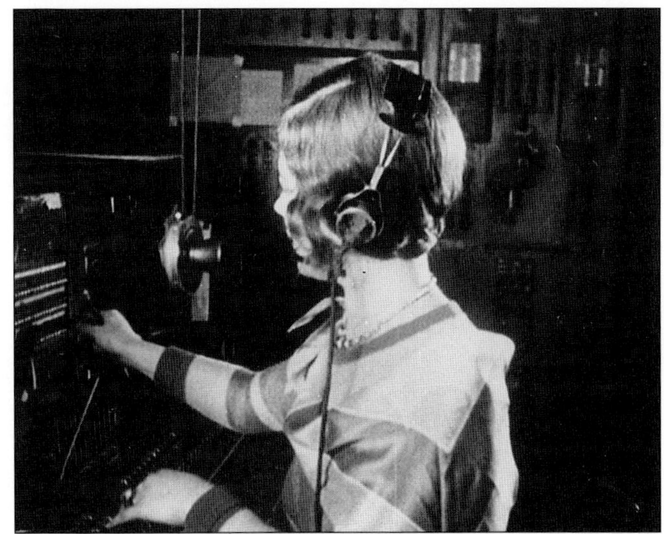
Bruce G. Moffat Collection

Telephone dispatching was used exclusively. These views, taken from a 1930's vintage newsreel show a train motorman conversing with the dispatcher who sat at a telephone switchboard.

the dual challenges of improving the system's public image and increasing traffic revenues. As a part of this effort a page-long article describing the system and its role in Loop-area commerce was prepared and published by the Chicago Association of Commerce in 1925. Titled "A Survey of Chicago," the narrative noted that in 1923, the system employed 500 persons and that 503,567 tons of merchandise, 28,421 tons of coal, and 71,263 cars of excavation and cinders had been handled. Also included was this table of operating statistics for four selected years:

Item	1910	1915	1920	1924
Area Served (Sq. Miles)	4	4	4	4
Investment in Property	Receivership	$28,163,928	$29,306,010	$29,409,829
Miles of Track	59.58	59.63	59.63	59.85
No. of Tunnel Connections:				
to Railroads	22	23	23	25
to Commercial Houses & Bldgs.	45	48	54	60
No. of Locomotives	128	131	130	146
No. of Tunnel Cars	3,000	2,979	2960	2,808
Tunnel Cars Handled:				
Public Stations	191,690	259,913	94,943	211,040
Railway Interchange	164,045	140,768	46,477	76,164
Commercial Houses	146,935	178,230	160,121	153,913
Excavation	47,595	20,009	8,483	34,673
Coal	8,514	16,122	4,993	9,052
Cinders	13,974	17,000	25,958	26,507
Total Cars Handled	572,753	632,042	340,975	511,349
Equivalent Vehicles Removed from Streets	1,381,965	1,343,787	799,754	1,111,760

There was no doubt that the tunnel system was a large business concern and that these statistics were meant to point out that management had made great strides in reversing the traffic declines of the late 'teens. There remained, however, much work to do as evidenced by the commercial house statistics which showed a continuing decline even though the number of building connections had actually increased. Unfortunately, the succeeding years would see traffic in all categories start to decline for a wide variety of reasons.

Bruce G. Moffat Collection

The Grand Avenue disposal station in 1928. This location is unrecognizable today.

The Grand Avenue Station

With the completion of the landfill operations in Grant and Burnham parks a new disposal site for heating ash generated by the system's many customers was needed. Accordingly, property was secured for a new disposal station to be operated by the Chicago Warehouse & Terminal Company. Located along the west bank of the Chicago River just south of Grand Avenue, the newest and last of the system's disposal stations opened for business on May 28, 1928.

Loaded cars were raised out of the tunnel using derricks and their cargo transloaded into scows for movement to the final disposal site. This station remained in operation until the end of all service in 1959. By that time, however, the scows had long since been replaced by railroad gondolas provided by the Chicago & North Western Railway.

Following abandonment of the system, the Grand Avenue site was used as a Department of Public Works storage yard. In 2000, the site was being redeveloped with luxury housing.

A New Franchise

The 30-year franchise originally granted in 1899 to the Illinois Telephone & Telegraph Company to build and operate its "conduit" system expired on February 19, 1929. Although the franchise had been amended on several occasions, the original expiration date had remained unchanged.

In January 1930, the city advertised for bids to lease and operate the tunnel system. The only bid received was from the Chicago Tunnel Company. The company and the city then sat down to hammer out a new franchise. The City Council created a special committee made up of members of the Committee on Local Transportation and the Committee on Gas, Oil and Electric Light to handle the negotiations.

In drafting the new franchise, many points of disagreement arose that slowed the process considerably. Perhaps the most serious point of disagreement was the issue of tunnel ownership. The city insisted on a strict interpretation of the amended franchise which gave the city uncontested ownership of all tunnels located under public streets, alleys and the Chicago

River upon the grant's expiration in 1929. In the end, the company had little choice but to yield to the will of the city on this point.

The company was also concerned about the rate of compensation it was required to pay the city under the terms of the old franchise, believing it to be too high. During the last ten years of the old grant the company had to pay 12% of its gross revenues while other utilities were reportedly paying an average of only 3% for the right to occupy the public way. Company officials claimed that this rate represented unjustified discrimination against a firm struggling for its very survival.

A third point of contention, and perhaps the most important of all in terms of having a lasting effect on the company's profitability, concerned the proposed condemnation and removal of tunnels for the construction of passenger subways beneath State Street and other thoroughfares (subway planning was being actively pushed at the time). Subway construction was likely to claim some of the company's most profitable trackage as well as isolate parts of the system or necessitate long detours. Also, some of the company's most profitable merchandise connections were located along State and Dearborn streets and would be lost if subways were constructed and new connecting tunnels not provided.

It was not until 1932 that all of the issues were resolved and incorporated into the proposed franchise ordinance and forwarded to the full council for its review. Approval came on July 19, 1932. The new grant gave the Chicago Tunnel Company permission to "use and occupy existing tunnels of the City of Chicago, together with extensions thereof, additions thereto and substitutions thereof." The new franchise was for

Bruce G. Moffat Collection

Portrait of Tunnel Company president Sherman W. Tracy published in 1932.

Chicago Sun-Times

Motorman James Comyns aboard a Baldwin in 1938. Note the "73" on the wall just above the headlight. This was part of a location numbering system. Unfortunately a listing of customers by number has not survived.

Bruce G. Moffat Collection

Two units from the company's dwindling fleet meet at the turnout from Public Station #3 on the four-rail joint trackage in the early 1950's.

a 30-year term and went into effect on September 16. The company also had to pay the city $710,000 as settlement on the delinquent franchise payments.

The franchise also affirmed the city's claim of ownership of all tunnels not located under private property, but did not include track, rolling stock, conduits, pumps or other equipment contained in the tunnels; these remained in Tunnel Company ownership. The city further reserved the right to order the removal or relocation of any tunnel it deemed necessary to allow for the construction of passenger subways or any other improvement that would require encroachment upon the tunnels. The company agreed to waive all claims arising from the loss or damage of its property and business in connection with such removal or encroachment. The city did agree, however, to reimburse the company up to $325,000 for costs incurred in altering, lowering or reconstructing tunnels or in building bypass tunnels in connection with the construction of the State Street Subway and $375,000 for like work in connection with the building of the Dearborn Street Subway.

The company was granted the use of the space under the tunnels (up to a depth of five feet below the base of the rail) for use as part of its system for the "installation, maintenance and operation of conduits, pneumatic tubes, pipes, wires, cables and other facilities and equipment for any public service."

New rates of compensation to the city were also fixed. These fees were based on the amount of freight handled through the tunnels: three cents per car of bulk freight (coal, ash, clay, excavated materials, etc.) and three cents per ton of package and parcel freight. These rates were based on tariffs in effect at the time and were to be changed proportionately with any alteration of the tariffs. A charge of six percent of the gross receipts for use of the tunnels for purposes other than the above was also established.

This new franchise represented a major redefinition of the relationship between the parties, with the company making major concessions relative to ownership and use of the tunnels in exchange for a more favorable franchise fee agreement.

Bruce G. Moffat

Some of the Tribune's *12" gauge paper cars survived into the 1970's.*

The Chicago Tribune

During their first three decades, the tunnels were used solely by trains of the Chicago Tunnel Company and its predecessors. This changed in 1934, when an agreement was made between the Tunnel Company and the Chicago Tribune allowing the newspaper "tracking rights" so it could move rolls of newsprint from its warehouse on North Water Street to its printing plant located in the basement of the Tribune Tower located about 1,400 feet away. The pact was signed on July 28, 1934, and gave the newspaper the right to

Bruce G. Moffat

This 1980 view shows the conveyor that replaced the 12" gauge cars for moving newsprint between the Tribune's river front warehouse and the pressroom.

Bruce G. Moffat

A look down the 20% grade from the Tribune Company's warehouse. In 1980 this ramp was still in place but unused.

use portions of the Austin Avenue and North Water Street tunnels between 10:00 p.m. and 6:00 a.m. when regular tunnel trains were not operating.

Until this time, there had existed no physical connection between the two tunnels even though they ran within inches of each other for 70 feet. To overcome this obstacle a 34-foot section of wall was removed so that a track connection could be made.

Instead of opting for tunnel-prototype two-foot gauge equipment, the Tribune chose to use small 4-wheel cars operating on 12" gauge track. The operating plan required the installation of a third running rail between the two existing ones, and a fourth rail between the track and the outside wall. This resulted in a double-track path for the Tribune's paper cars while preserving the two-foot gauge track for continued use by the Tunnel Company during the day.

Access from the Tribune's warehouse to the tunnel was provided by a two-track ramp having a 20 percent grade. This unusually steep incline allowed a loaded car to gain sufficient momentum to propel it all the way into the newspaper's basement, eliminating the need for locomotive power. By 1980, the gravity system had been replaced by a conveyor which used the old elevator shaft in former Public Station #3 (adjacent to the Tribune's warehouse and dock) to reach the tunnel.

The Tribune subsequently constructed a new multimillion-dollar printing plant on the west bank of the Chicago River just south of Chicago Avenue. Although the new facility became operational in 1981, the old printing facilities in the Tribune Tower (and the tunnel) remained in use through 1982. Since that time, the former public station and the Tribune's dock and storage facilities on North Water Street have been demolished as part of a major urban renewal effort that has seen the area evolve from an industrial area into a high-density residential area and tourist destination.

The Tribune had at one time considered using other portions of the tunnel system to move materials between Tribune Tower and the new printing plant (which lies in part on the site of an old Chicago & North Western Railway freight facility once served by tunnel). This plan was rejected because of the prohibitive costs associated with rehabilitation of the connecting bores.

Bruce G. Moffat Collection

Needing a photograph to illustrate a news article about the March 1937 strike, a Chicago Tribune *photographer recorded this scene of an "inspector" checking idled cars. The location is the junction of the Hubbard (Austin) and North Water tunnels-a short distance from the newspaper's offices. The "inspector" is most likely another newspaper employee. Note the four-rail trackage shared by Tunnel and Tribune cars.*

The Strike

Like its standard-gauge counterparts, the tunnel system experienced occasional labor difficulties. These disputes usually attracted little press coverage because of the limited scope of the company's operations, the availability of alternative truck transportation, and the "out of sight - out of mind" nature of the tunnel system. All of this was to change in 1937.

Chicago Tunnel Company motormen, along with the freight handlers of the Chicago Warehouse & Terminal Company, were represented by Tunnel Lodge #125 of the Brotherhood of Railway & Steamship Clerks, Freight Handlers, Express & Station Employees. On the afternoon of March 9, 1937, the members of Lodge #125 struck the two companies over alleged contract violations.

The major grievance concerned the amount of freight traffic being diverted from the tunnel trains to the trucks of the affiliated Chicago Tunnel Transport Company. Diversion of such traffic meant that the Tunnel Company avoided city franchise payments because amount due was based on the amount of traffic handled through the tunnels. No compensation was due for cargo moved by truck.

The Brotherhood charged that more than the agreed 50% of total traffic handled was being moved by truck, resulting in the loss of 35 motormen's jobs since their existing contract went into effect on December 1, 1936. Seventeen other grievances were also cited, and the union further demanded that all future disputes be settled by arbitration.

Instead of setting up a picket line, the employees staged a sit-in at their work locations, refusing to leave. The freight handlers went so far as to barricade the doors to the Public Stations and the Grand Avenue disposal facility.

The strike ended peacefully on March 11. Management agreed to reinstate ten of the 35 motormen, and further consented to allow the National Adjustment Board (set up by the Railway Labor Act, to which the Tunnel Company was subject) to arbitrate any future disputes.

The Subway Cometh

While a very limited amount of trackage had been abandoned prior to 1940, it was not until 1941 that wholesale tunnel closures would occur. These reductions would come as a direct result of the city's program to build a system of passenger subways in the Loop area.

Constructed generally to the same depth as the freight tunnels, the subways caused the removal or abandonment (due to some segments becoming physically isolated) of 5.94 miles of track and tunnel during the first six years of construction, with some additional losses in subsequent years.

Those tunnels removed to permit construction were:

- State Street, from Illinois Street to south of Roosevelt Road
- Milwaukee Avenue, from Lake Street to north of Fulton Street
- Lake Street, from Milwaukee Avenue to just west of Dearborn Street
- Dearborn Street, from south of Lake Street to Congress Parkway.

Various other tunnels that crossed the path of the as-yet unbuilt Congress Parkway and Eisenhower Expressway to be used by the future Dearborn subway were also severed.

Only two tunnels that crossed the path of the advancing construction went undisturbed: Market Street (now Wacker Drive) where it crossed Lake Street and Franklin Street at Congress Parkway. Both of these were above the elevation of the subway tubes.

Also emerging apparently more-or-less intact was the Milwaukee Avenue tunnel from a point a half-block north of Fulton to Kinzie. In this area, the twin subway tubes separated and passed by the tunnel on either side to avoid the supports for the old Milwaukee Avenue viaduct which spanned the railroad tracks leading into Union Station. However, all connections to this bore were severed by the subway construction, resulting in its becoming totally isolated.

Other tunnels isolated by the subway construction, but otherwise surviving intact included those beneath east-west streets from Lake to Quincy streets (being interrupted where they crossed State and Dearborn streets), and the Wacker Drive tunnel from State to LaSalle streets, as well as connecting bores

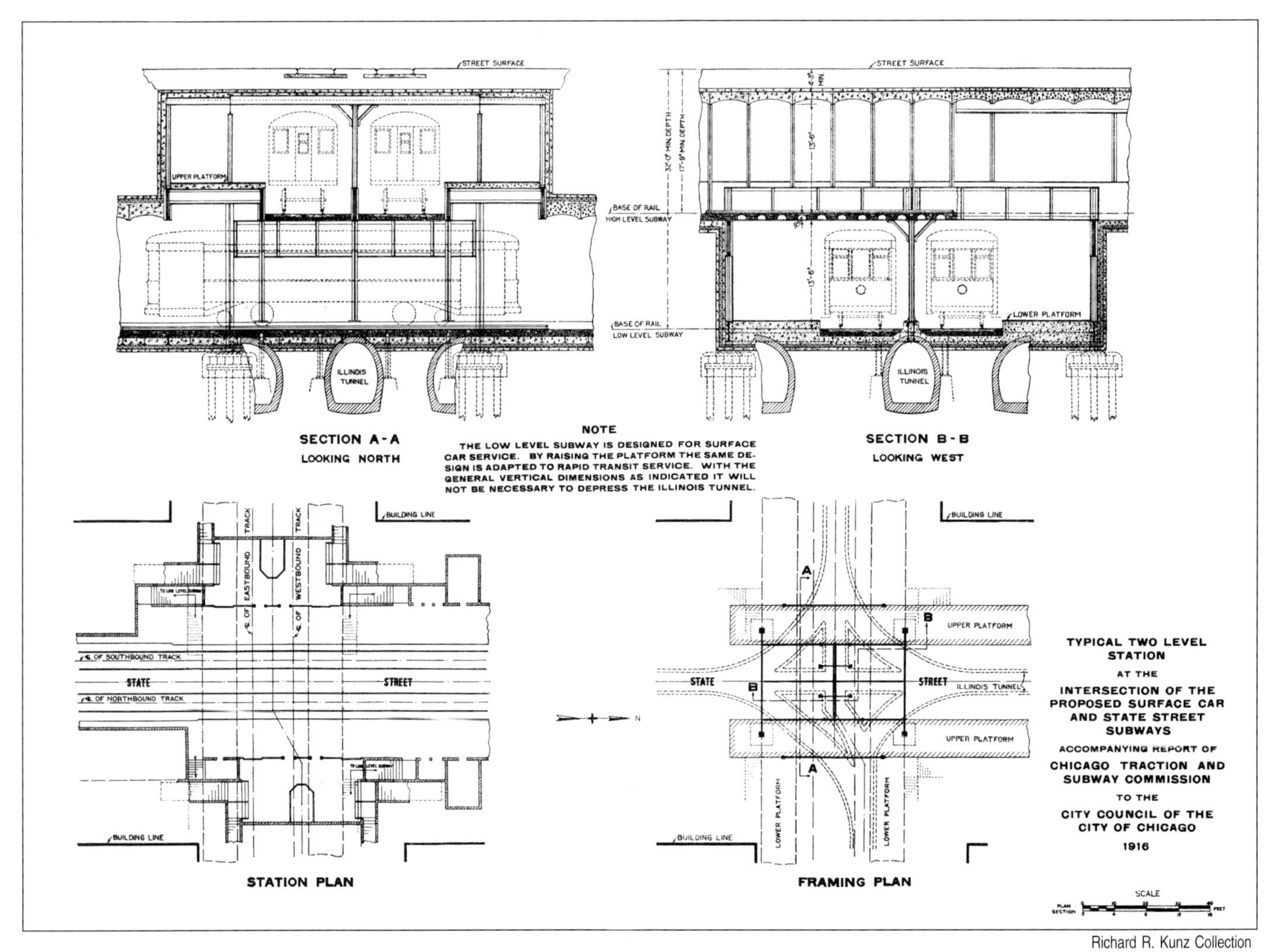

Richard R. Kunz Collection

This 1916 proposal for a downtown subway system for streetcars and rapid transit trains managed to avoid displacement of the underlying freight tunnels – but just barely. By the mid-1930's, subway planners made little or no effort to spare tunnel trackage.

Bruce G. Moffat Collection

In 1923, Mayor William E. Dever (center) visited the tunnels to proclaim his confidence that subway construction would begin in the near future. Nearly two decades would pass, however, before work finally began on the State Street subway.

Construction of the State Street subway, June 13, 1940. Note the temporary two-foot-gauge construction tracks and one of the Tunnel Company's ash cars used to haul away spoil in the downtown area.

Chicago Transit Authority

under Dearborn Street and Clark Street north of Lake Street. Interestingly, the isolated Madison and Quincy tunnels actually opened into the State and Dearborn subways at each end (apparently for subway construction access) but were not sealed until after the 1992 Loop flood.

Unfortunately for the company, much of the trackage that was removed was among the busiest and most profitable on the system, serving numerous large department stores and office buildings along State and Dearborn streets. Also lost was important connecting trackage along Lake Street and Milwaukee Avenue.

It is ironic to note that, during the initial construction of the freight tunnel system, the promoters insisted that the tunnels were being built to a depth that would allow for the construction of passenger subways above them at some future time. Except where these tubes would have had to descend to cross under the Chicago River, the freight tunnels would not have been interrupted. Several early subway plans called for relatively shallow subways to be built immediately above the tunnels. Even some later plans that called for double deck rapid transit and streetcar subways made provisions for the tunnels. As late as 1932, a plan developed by the city's Board of Local Improvements offered a proposal that would have allowed the freight tunnels to coexist with the proposed State Street Subway.

In 1939, the company proposed construction of

To get around the subways several bypasses were built. At left in this 1992 view looking west on Randolph from Wabash, is the beginning of the bypass tunnel that passes beneath the State and Dearborn subways.

Bruce G. Moffat

eleven bypass bores to connect the major parts of the system that were to be cut off by the subways. These bores were to be located:

1) Under subway in State Street at Illinois Street.
2) Under subway in State Street at Hubbard Street.
3) Under subways in State Street at Harrison Street.
4) Under subways in State and Dearborn Streets at Randolph Street.
5) Under subways in State and Dearborn Streets at Washington Street.
6) Under subways in State and Dearborn Streets at Adams Street.
7) A "northwest" bypass near Roosevelt Road and Wabash Avenue (crossing subway in State Street).
8) Under subway in Lake Street at Franklin Street.
9) Under subway in Lake Street at Wells Street.
10) Under subway in Milwaukee Avenue at Jefferson Street.
11) Over subway in Milwaukee Avenue at Canal Street.

According to its franchise, these bypasses were to be built by the company with construction costs to be reimbursed by the city from a special fund specifically set up for that purpose. Ultimately, it was the city that constructed the bypasses but only at six locations (a seventh crossing required only limited modifications to remain in use). These were:

1) On Illinois Street, crossing under the subway at State.
2) On Randolph Street, crossing under the State and Dearborn Street subways.
3) On Franklin Street, crossing under the subway at Lake Street.
4) On Jefferson Street, crossing over the subway at Milwaukee Avenue.
5) On Sherman Street, crossing over the subway at Congress Parkway.
6) On Canal Street, crossing over the subway at Congress Parkway.

These last two bypasses were built in the early 1950s, and saw little if any use before being abandoned when freight activity in the southern portion of the system evaporated. The Sherman Street bypass was in service such a short period time that the pouring of concrete to cover the ties had not yet occurred when service in this area was discontinued.

The seventh bypass was not one at all, but rather involved the rebuilding of the Roosevelt Road tunnel where it intersected the old State Street bore. At this location, the bottom of the subway tube (which is gradually rising to surface level at this point) was at about ceiling level of the freight tunnel. This condition necessitated only the removal of the single-track grand union, the filling in of the State Street bore and the installation of a new tunnel roof for the surviving segment on Roosevelt to allow continued east-west operation.

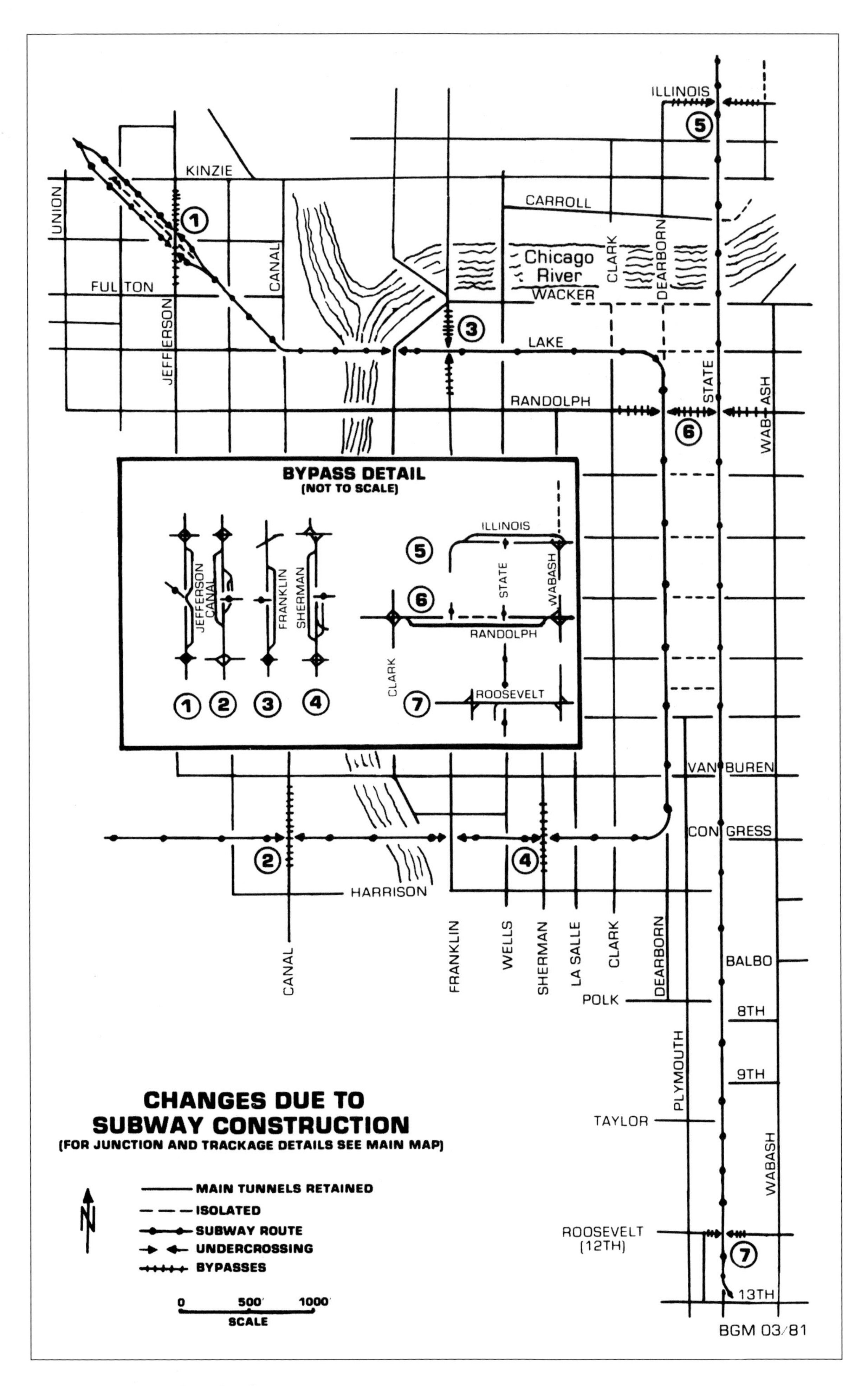
ILLINOIS
KINZIE
CARROLL
UNION
CANAL
FULTON
JEFFERSON
Chicago River
WACKER
CLARK
DEARBORN
LAKE
STATE
RANDOLPH
WABASH
BYPASS DETAIL
(NOT TO SCALE)
JEFFERSON
CANAL
FRANKLIN
SHERMAN
ILLINOIS
STATE
WABASH
RANDOLPH
CLARK
ROOSEVELT
VAN BUREN
CONGRESS
HARRISON
CANAL
FRANKLIN
WELLS
SHERMAN
LA SALLE
CLARK
DEARBORN
POLK
BALBO
8TH
9TH
PLYMOUTH
TAYLOR
WABASH
ROOSEVELT (12TH)
13TH
CHANGES DUE TO SUBWAY CONSTRUCTION
(FOR JUNCTION AND TRACKAGE DETAILS SEE MAIN MAP)
MAIN TUNNELS RETAINED
ISOLATED
SUBWAY ROUTE
UNDERCROSSING
BYPASSES
N
0
500'
1000'
SCALE
BGM 03/81

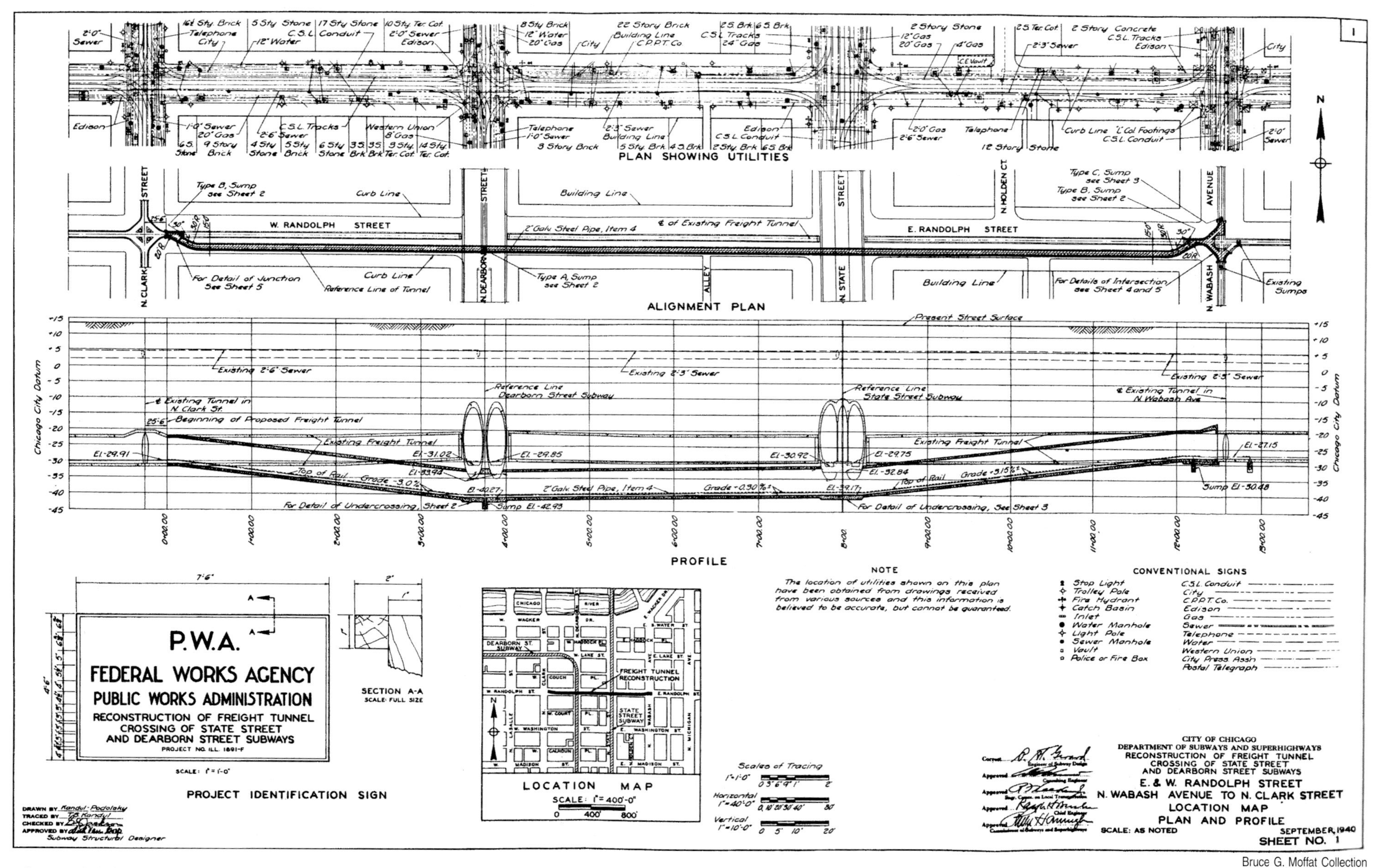

Bruce G. Moffat Collection

The Randolph by-pass crossed beneath both the State and Dearborn subways.

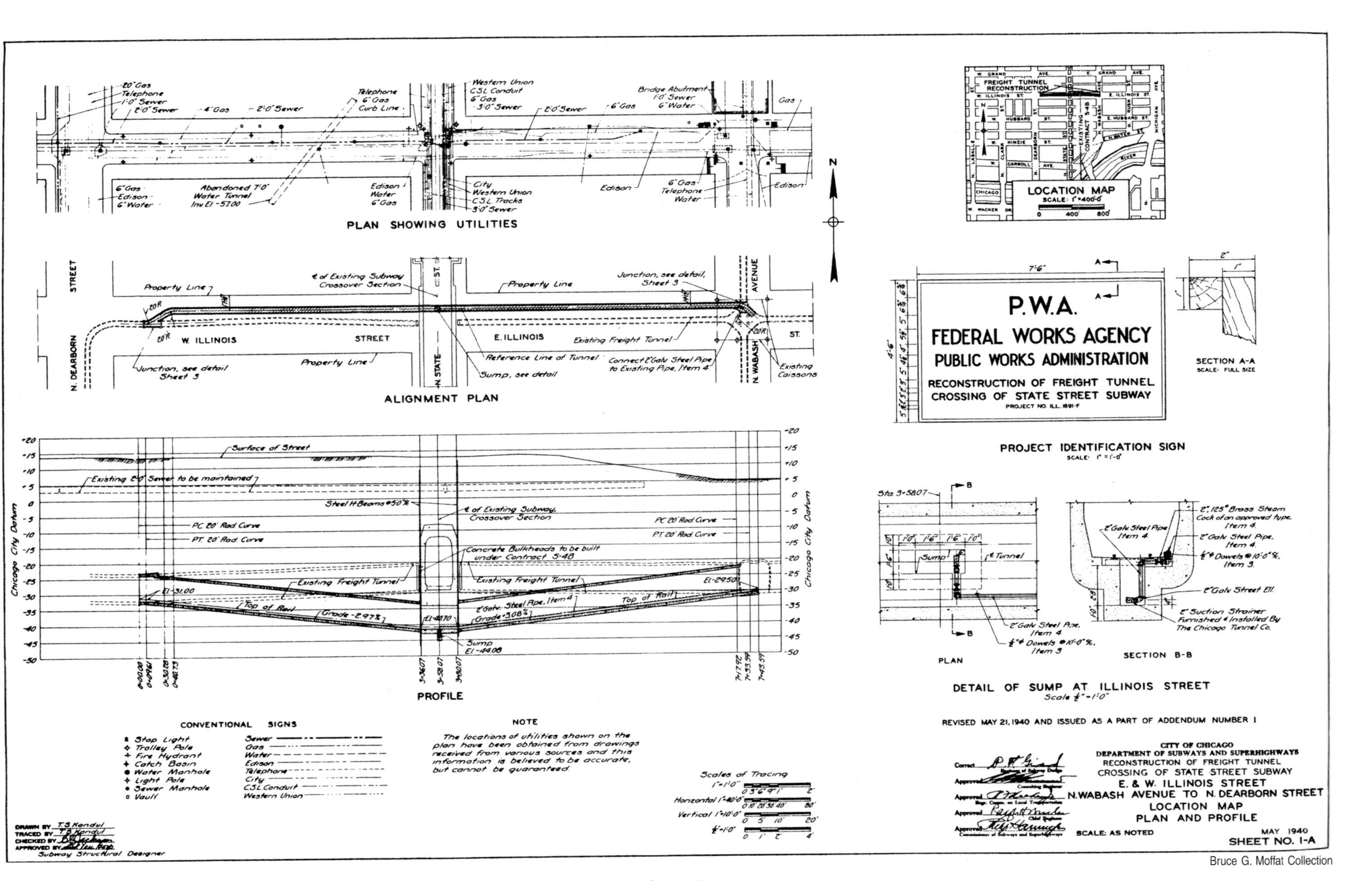

Bruce G. Moffat Collection

The Illinois by-pass allowed continued access to the river front warehouses located east of State Street.

Measuring the Tunnel

The various annual reports, regulatory agency filings, and publications prepared by the company frequently gave conflicting figures on the amount of trackage (and tunnels) built and operated over the years. Promotional literature issued during the years that the system was at its greatest extent gave lengths varying from 61.96 to 65 miles.

A 1922 report by the Bureau of the Census fixed the total trackage at only 59.63 miles, while a 1949 examination of the Tunnel Company's own atlas by the Chicago Plan Commission could only document the existence of 58.85 miles of track prior to the start of subway construction.

Yearly reports filed with the Interstate Commerce Commission (ICC) by both the Chicago Tunnel Company and the Chicago Warehouse & Terminal Company are perhaps the most authoritative and consistent figures available on the subject. A sampling of these data is given below. The lengths listed for sidings and the turnouts leading to them were reported by both companies since CW&T leased this trackage from its sister firm and both companies were under ICC jurisdiction.

Year	Chicago Tunnel Co. Main Miles	Jointly Reported Miles	Total
1915	52.00	(not reported)	52.00
1916	51.51	8.12*	59.13
1922	51.51	8.12*	59.13
1933	48.98	10.18	59.16
1940	38.92#	8.75	47.67
1959	38.54##	8.63	47.17

* Includes .5 miles of main track reported by CW&T
# Reduction due to State and Dearborn subway construction
Reduction due to extension of Dearborn subway under Congress Street

Tunnel vs. City

On October 17, 1943, the Chicago Rapid Transit Company inaugurated passenger service through the city-owned State Street Subway. Although the important State Street tunnel was now irretrievably lost, the Tunnel Company had (quite belatedly) just begun to fight. It was in that same year that the company filed suit in the Superior Court of Cook County against the city.

In its complaint, the company charged that the city had not lived up to the terms of the July 19, 1932 franchise ordinance, having "...entered upon, encroached, seized, taken, destroyed and rendered useless to the plaintiff various parts and lengths of its aforesaid 'tunnel system'...thereby making it necessary to alter, lower, relocate and reconstruct 'tunnels' and parts thereof and to construct connecting tunnels as a substitute for the 'tunnels' which the defendant removed, intercepted or destroyed by the construction of said system of subways..."

The company further charged that the city had refused to set aside the necessary funds required to pay for the construction of the eleven bypasses that the company had deemed necessary to continue operations in an efficient manner. It was alleged that the city only appropriated funds sufficient to pay for the building of four such bypasses (up to that time). With only four by-passes available, tunnel trains had to take lengthy detours to reach customers.

Interestingly enough, the city responded that it did not have the power to grant a lease of the tunnels, even though it had in fact done so under the terms of the 1932 franchise. Further, since no appropriation for tunnel relocation work had been made by the city prior to the adoption of that franchise ordinance, the issue was null and void; the plaintiff Tunnel Company was not entitled to recover damages.

The city maintained that the company had refused to relocate or reconstruct any tunnels interrupted by subway construction at its own expense and then petition for reimbursement as provided for in the franchise. Instead the company had requested that the city undertake the construction resulting in the waiving of all claims for liability. The city also pointed out that a good portion of the company's traffic had been diverted to trucks to reduce costs.

For the next 14 years, an almost endless series of continuances were granted by the court with no progress being made towards its resolution. The last continuance was granted on June 27, 1957. The case was once again put on the court's calendar to be heard, this time on September 23. At that hearing, the company abandoned further legal action, having finally realized the futility of continuing to press ahead. By this time, what few trains were still running were only hauling ash; all other traffic having been lost. It was clear that the end was near.

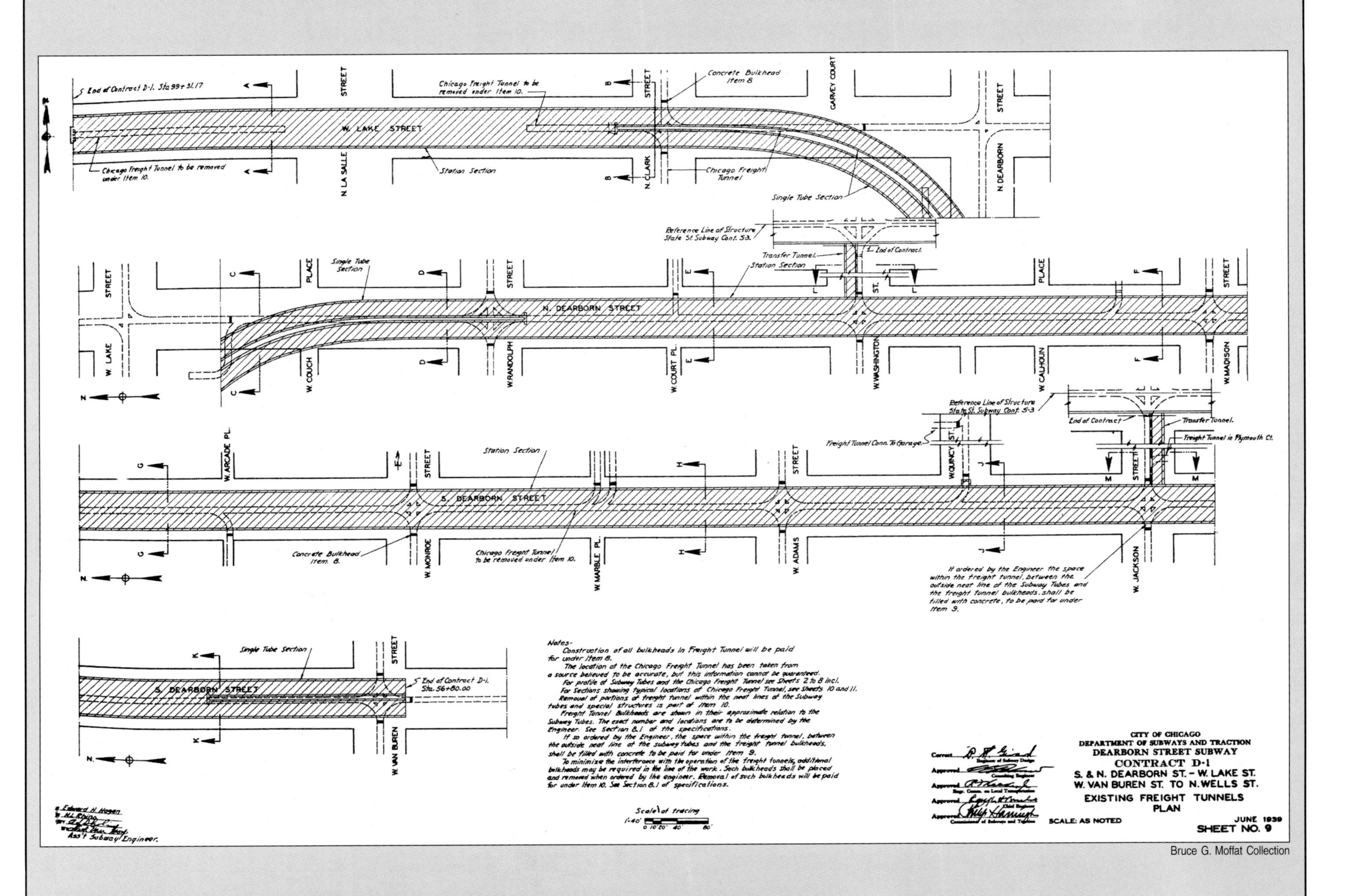

Construction of the Dearborn Subway (begun before World War II but not opened until 1951) obliterated the important tunnels beneath Lake and Dearborn as well as severing a number of important east-west routes.

Chicago Transit Authority

The first train in the State Street subway, April 2, 1943. Regular operations would not commence for another six months.

Competition and Declining Traffic

Built to handle LCL (less-than-carload) freight and retail delivery of coal, it is not surprising that the steady loss of this business (especially after 1936) would eventually lead to the system's abandonment.

It was in 1939, that the railroads serving Chicago inaugurated the free pickup and delivery of LCL shipments. The consequences of that action were dramatic. In 1933, tunnel trains handled 404,948 tons of cargo. By 1939, loadings had dropped to only 177,355 tons, accounting for about 6.4% of all such traffic being handled in the Chicago area. Except for a brief resurgence during World War II, this traffic continued to slip away.

Another factor was the gradual relocation of railroad freight houses to outlying areas. This trend began in the late 1920s when the Chicago & North Western Railway opened a new freight house at its Proviso Yard more than 12 miles away. Although the North Western continued to handle some freight at its old downtown location, one by one the other railroads followed suit, diverting as much of their LCL activities out of the downtown area as possible.

The increased use of trucks in general further diminished tunnel revenues. Unhampered by narrow tunnel clearances or limited operating areas, trucks proved to be an inexpensive alternative – especially for coal deliveries, where gravity instead of costly conveyors could be used to fill building coal bins. In addition, the gradual conversion from coal to oil and gas heating was picking up momentum, which meant the loss of ash traffic as well.

The large downtown department stores were increasingly relying on trucks, which could be loaded right at the store and for direct delivery to the

Bruce G. Moffat Collection

An electric "mule" is used to position a string of merchandise cars at what appears to be the Steele-Wedeles Company's river front warehouse in 1928.

Meanwhile forty feet below, another load of small packages is taken on a short train ride through the tunnels.

Bruce G. Moffat Collection

customer. This not only saved time, but also money, because there was no need to reload cargo at a public station. Trucks were also better equipped to handle the many large and bulky articles that could not fit through the tunnels. Although the Tunnel Company's own trucking firm, the Chicago Tunnel Transport Company (formed in 1927), was initially successful, its profitability eroded during the 1950s.

More Corporate Changes

On April 8, 1941, the Chicago Warehouse & Terminal Company changed its name to the Chicago Tunnel Terminal Company of Illinois. This new corporate title remained in use for less than one month before it was changed to the simpler Chicago Tunnel Terminal Company (effective May 6); its nomenclature differing from its parent only by the substitution of "Company" for "Corporation." Aside from a minor change in the wording of its charter, the rationale behind these alterations is unclear, as the relationship between the various affiliated firms remained unchanged.

The postwar years were not good ones for the system. The loss of profitable trackage to subway construction, truck competition, high operating costs,

Bruce G. Moffat Collection

An "inspection party" disembarks in the basement of a Loop building in 1937. Two of the individuals have leaned slightly forward to avoid accidental contact with the live trolley wire above the tracks.

Bruce G. Moffat Collection

Deferred maintenance was apparently the order of the day in 1935 judging by the appearance of this locomotive.

declining traffic caused by customers closing or moving away, and the cumulative effects of years of deferred maintenance all began to take their toll.

In 1948, the associated companies were forced to vacate their building at 754 W. Jackson Boulevard to make way for the eventual construction of the Kennedy Expressway. Not only was the office space lost, but also the space occupied by the main shop facility and the public station.

The offices were relocated to 340 West Harrison Street, where they remained until 1951, at which time they were again relocated to the Brooks Building at 223 W. Jackson Boulevard. This move was a re-entry to familiar quarters, as the Illinois Tunnel Company had once maintained a telephone exchange there.

7 Projecting an Image from "Forty Feet Below"

The "out of sight – out of mind" nature of the tunnel system's railway operations meant that management had to be more proactive than traditional railroads to keep their company's name before the public and to maintain a positive public image. This was especially true in the years prior to the renegotiation of the system's franchise in 1932. The tumult that surrounded that process marked the start of a decline in the system's promotional activities. Oddly enough, the tunnels would later serve as inspiration for a radio drama and several motion pictures.

Best Foot Forward

From the beginning of operations in 1904 through the 1920s, the tunnel system relied on print media to get its story out. Generally, these articles, which appeared in newspapers, popular magazines, and even railroad industry trade journals, tended to describe tunnel operations in glowing terms. Even though they included some statistics about the number of cars and locomotives in the fleet and businesses served, the level of detail seldom moved far beyond the general, making up for it through the use of flowery

International Film Service/National Geographic Society Collection

During the early years carefully arranged publicity events were often used to garner positive publicity for the system. This jovial group is apparently on a tour of the newly constructed mail handling facilities in 1906. Interestingly, this photograph appeared in a 1919 National Geographic article about Chicago.

The Colorful World "Forty Feet Below"

Bruce G. Moffat Collection

This "colorized" 1904 view of one of the test trains was issued by Chicago postcard publisher V. O. Hammon Company.

Bruce G. Moffat Collection

Retouching images for postcard use was sometimes so complete that the resulting image was now a painting rather than a photograph. The street signs identify the location as Dearborn and Van Buren

Bruce G. Moffat Collection

This very unusual postcard shows not only the lower levels of Marshall Field's department store on State Street, but also the freight tunnel and the various strata of dirt, clay and sand.

Because color photographic emulsions did not become generally available until the 1930's, color images of the tunnel trains are almost non-existent. Fortunately, the Illinois Tunnel Company had arranged with several postcard publishers to publish and distribute several views in the 1904-1910 period. The original black and white images were hand colored by artists in Germany, who had no personal familiarity with the subject matter. As a result, the choice of colors was left to their discretion unless specific instructions had been given. The "colorized" images were then returned to the United States where they were printed and sold.

Bruce G. Moffat Collection

A head-on view of one of the trolley locomotives. As was frequently the case, the colors used on the locomotive are not authentic. At one point the locomotives were painted an olive green.

Bruce G. Moffat Collection

Perhaps the most common of all tunnel post cards is this one showing one of the Morgan third rail locomotives facing southbound in the Franklin trunk tunnel approaching Madison.

Bruce G. Moffat Collection

Far less common is this adaptation commissioned by the Selz Shoe Company.

prose. The following article that appeared in the May 1923 issue of *Electric Traction* is a case in point:

Chicago Has A Subway

Extensive Electric Railway System of the Chicago Tunnel Company Underneath Loop Streets Rivals Famous Underground Systems of Paris.

It probably would be difficult for the average citizen to imagine Chicago's already congested Loop streets bearing the burden of 5,000 additional wagon and motor truck movements each day, yet that is exactly what they would have to do if it were not for the Chicago Tunnel Company's sixty-four mile electric railway system which is operating day and night forty feet below the surface of the ground.

At every minute during the day and night one might stand at a given point in this underground network and see a seven car train hurry by, laden with dry goods for a State Street department store or a ten car train shuttling through with coal for a Clark Street hotel or another train rumbling along for the lake front with earth excavated from a new loop building site.

Without this underground system of hand-ling freight refuse, Chicago's traffic problem would give pedestrians and politicians and surface traction operators a good deal more to worry about than at present, which seems to be enough.

Practically every block inside the Loop district of Chicago is surrounded by the tracks of this extensive underground freight system. This means that it is possible for every commercial house and hotel in the loop to have direct connection with all the railroads and with one another entirely free from all traffic delays.

The tracks of this underground electric freight railway also extend outside of the Loop as far south as Fifteenth street, north to Superior street, west to Halsted street and east to the lake. Thus warehouses and shippers outside of the Loop are in direct uninterrupted connection with those inside the "dread square mile" as Chicago's congested Loop district is often called...

Probably not one in a hundred thousand of Chicago's populace know that such a subway even exists and still fewer know that this system has been in no small part responsible for the practical construction of Chicago's lake front "City Beautiful" plan.

Hundreds of thousands of cubic yards of excavated earth, refuse and ashes have been hauled from the loop through the tunnels to the water's edge and there made to push back the inroads of mighty Lake Michigan and to form the foundation of Grant Park and the new Field Columbian Museum.

This article also included "A typical statement showing the annual traffic handled" but without specifying the year:

At public freight stations		
275,218 tons		
Industrial	206,438	"
Between railroads	129,093	"
Coal	57,825	"
Total:	668,574	"
Excavation	160,342	cu. yds.
Cinders	58,978	" "
Total:	219,320	" "

Besides these "news" articles, the company's copy writers created a number of promotional booklets and pamphlets that described the system and extolled the invaluable role it played in the commerce of the nation's second largest city. Geared mainly towards prospective shippers, these lavishly illustrated publications with titles such as *Lifting the Lid in the Loop* and *What the Freight Tunnels Mean to Chicago* recounted in the same glowing but superficial prose how the little trains efficiently served the Loop's major buildings and expedited the movement of freight. Of course no retelling was complete without a tabulation of the fleet size, number of building connections, and miles of track. Sometimes a capsule history of the system's construction was included. Most of these publications also included a copy of the system map.

Speeches by company officials were also given to local civic and trade associations. Typical of these presentations was one made to a local chapter of the American Electric Railway Association, as reported in

Right: *Interesting statistics from a 1928 promotional booklet.*

General Information

PHYSICAL

Size of Tunnel	6′ x 7′ 6″	Lights in Tunnel	3,800
Miles of Tunnel equipped with 24 inch gauge track and trolley	61.96	Telephones in Tunnel	266
Intersections	734	Railroad Connections	49
Power: Electric, D. C. 250 volts (4 substations and 11 sections)		Universal Public Stations	4
Elevators	96	Private Merchandise Connections	26
Sumps	63	Coal and Cinder Connections	40
Pumps (large) connected with pumps	540	Cinder Connections	16
		Coal Receiving Stations	3
		Average Distance Below Street Level	40′

LOCOMOTIVE AND CAR EQUIPMENT

Electric Locomotives, 150

Total Cars, 3,304 (Merchandise, 2,693; Coal, 151; Excavation and Cinder, 400; Company Service, 60)

TRAFFIC HANDLED DURING A TYPICAL YEAR

MERCHANDISE	CARS	TONS
Public Stations	290,276	318,873
Commercial Houses	169,895	148,731
Railroad Interchange	129,590	133,368
Total Merchandise	589,761	600,972
Coal	16,414	57,440

Cinders 28,095 cars of 3½ yards capacity

Excavation and Refuse 74,084 cars of 3½ yards capacity

Average Number of Employes . . . 580

Average Annual Payroll . . . $957,873.72

PRIVATE TUNNEL CONNECTIONS FOR MERCHANDISE

Carson, Pirie, Scott & Co.
Durand & Kasper Co. (old building)
Durand-McNeil-Horner Co.
Eckhart Building (165 N. Wacker Dr.)
Franklin MacVeagh & Co.
Kraft-Phenix Cheese Co.
Hibbard, Spencer, Bartlett & Co.
Mandel Brothers (Warehouse)
Marshall Field & Co. (Warehouse)
Monarch Refrigerating Co.
National Candy Co.
Paper Mills Co.
Reid Murdoch & Co.
Selz Schwab & Co.
John Sexton & Co.
Sprague Warner & Co.
Steele Wedeles & Co.
United Fig & Date Co.

(*Note:*) *Cinders can be handled from all of these connections*

RETAIL STORE CONNECTIONS

Boston Store, The
Carson, Pirie, Scott & Co.
Davis Store, The
Fair, The
Mandel Brothers
Marshall Field & Co.
Woolworth Store

PUBLIC WAREHOUSES

C. & A. Terminal Warehouse Company (C. & A. R. R.)
Central Cold Storage Company
Crooks Terminal Warehouse (C. B. & Q. R. R.)
Currier Lee Warehouse Company
Great Western Warehouse (C. G. W. R. R.)
North Pier Terminal Company
Railway Terminal Warehouse Co.
Soo Terminal Warehouse Co.
Tooker Storage & Forwarding Company
Western Warehousing Company (Penn. R. R.)

UNIVERSAL PUBLIC STATIONS

No. 1—746 W. Quincy St.
No. 2—566 Kingsbury St.
No. 3—No. Water & Seneca Sts.
No. 4—Roosevelt Rd. & Canal St.

Approximately 2,000 shippers use these stations

COAL RECEIVING STATIONS

C. & E. I. R. R. (14th Street)
Crerar Clinch Coal Co. (Foot of So. Water St.)
Illinois Central R. R. (16th Street)

DISPOSAL STATION

Grand Avenue and North Branch of the River.

COAL AND CINDER CONNECTIONS WITH BUILDINGS

*Ashland Building
Boston Store
*Builders' Bldg.
Carson, Pirie, Scott & Co., Retail
*Central Police Station
*Century Bldg.
City Hall Bldg.
County Bldg.
Continental National Bank Bldg.
Conway Bldg.
*Durand, McNeil & Horner
Edison Bldg., The
*Federal Bldg.
Federal Reserve Bank
Field Museum
*First National Bank Bldg.
Foreman Bank Bldg.
Harris Trust Bank Bldg.
Heyworth Bldg.
*Hibbard, Spencer, Bartlett & Co.
Illinois Merchants Trust Co.
Insurance Exchange Bldg.
Kimball Bldg.
La Salle Hotel
Lytton Bldg.
Majestic Theater
Mallers Bldg.
Mandel Bros., Retail
Marshall Field & Co., Retail
Marshall Field & Co., Men's Store
McCormick Bldg.
Standard Trust Bldg.
Morrison Hotel
*Morton Bldg.
*National Republic Bank Bldg.
North American Bldg.
Otis Bldg.
Palmer House
Peoples' Gas Bldg.
Pittsfield Bldg.
Railway Exchange Bldg.
*Reid, Murdoch & Co.
Roanoke Bldg.
*Roosevelt Theater
*Selwyn-Harris Theaters
*Selz-Schwab & Co.
*Sexton & Co., John
Sherman House
State Bank of Chicago
Stevens Hotel
Straus Bldg.
*Transportation Bldg.
Tribune Tower Bldg.
Utilities Building
Woolworth Bldg.
*Wrigley Bldg., Annex

*(*Cinders only*)

Page Thirty-two

Bruce G. Moffat Collection

the October 1928 issue of the Chicago Rapid Transit Company's employee magazine, *The High Line*:

Interesting Story of Chicago Tunnel Railway Told by Speaker

Regular monthly meetings of Chicago Rapid Transit Company Section No. 6, A.E.R.A. were resumed on the evening of September 18. The "headliner" on the program ushering in the new season of activities was J. H. Burke, Traffic Manager of the Chicago Warehouse and Tunnel [sic] Company, who told the story of Chicago's under-ground railway of 62 miles in length, which handles an average of approximately 600,972 tons of freight each year.

This tunnel railroad, having connections with practically every large wholesale and retail store, office building, hotel, public warehouse, railroad yard and industrial enterprise in the Loop and nearby area, does an annual volume of business totaling several millions of dollars, Mr. Burke revealed.

Relieves Street Traffic

Operated under dispatcher control and in all physical aspects resembling an electric railroad, this "subway" system has 150 electric locomotives and 3,304 cars. The size of the tunnel through which the road operates is six feet by seven feet six inches. The bulk of business is in transporting merchandise between commercial houses and shipping points, thus relieving the city streets of a majority of heavy delivery traffic.

Another interesting detail of service mentioned by Mr. Burke was that relating to the sale of air to Loop theaters and hotels for cooling and ventilating purposes. The temperature of the air in the tunnel is about 55 degrees throughout the year. It is also 20 per cent purer than surface air, he declared. Loop institutions purchase the air, which is distributed through ventilating systems attached to huge intake pipes installed in the tunnel.

Although these presentations seldom dealt with the "nuts and bolts" aspects of the operation, they were nonetheless interesting and reflected the positive public relations posture of the system's management during this period.

Newsreel cameras visited the tunnels on several occasions during the 1920s and 1930s. Although some of this footage does survive, story details do not. About all that can be deduced is that these filmings were for general interest pieces rather than to illustrate any "hard news" story.

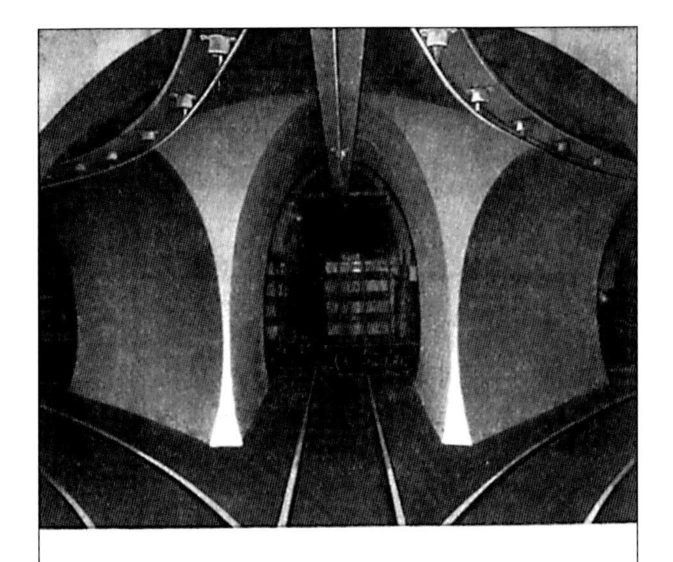

Large and small shippers receive same service

SHIP VIA TUNNEL

Here's a view of what I saw 40 feet under the ground Saturday afternoon 1/21/28.

CHICAGO WAREHOUSE AND TERMINAL CO.
754 WEST JACKSON BOULEVARD

TELEPHONE HAYMARKET 6300 — TRAFFIC DEPARTMENT

Bruce G. Moffat Collection

Promotional card given to potential shippers by the Chicago Warehouse & Terminal Company's traffic agent. It appears that someone had an enjoyable tour of the railroad "forty feet below."

Bruce G. Moffat Collection

A workman on the Selz Schwab & Co.'s loading dock beneath Monroe Street west of Market Street loads crates on a waiting tunnel car circa 1910.

Issued in 1928, this booklet relied on an aerial photograph to illustrate the system's reach. Bruce G. Moffat Collection

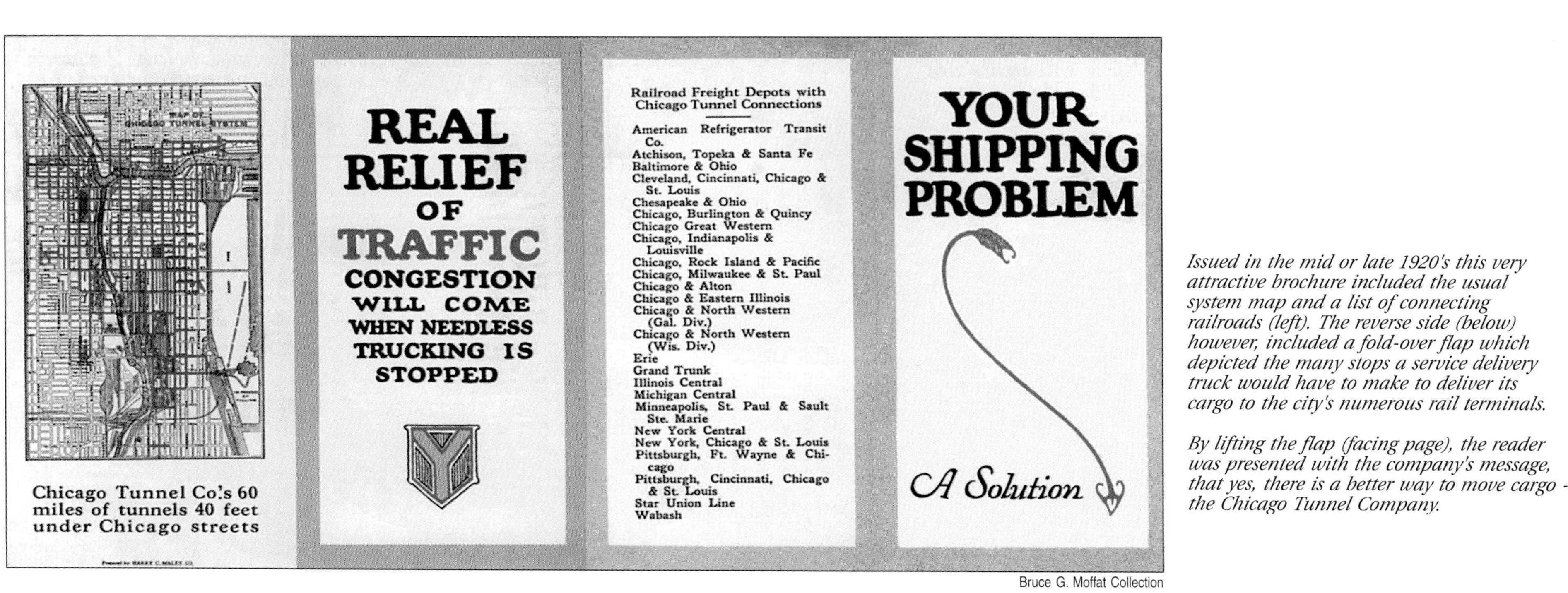

Bruce G. Moffat Collection

Issued in the mid or late 1920's this very attractive brochure included the usual system map and a list of connecting railroads (left). The reverse side (below) however, included a fold-over flap which depicted the many stops a service delivery truck would have to make to deliver its cargo to the city's numerous rail terminals.

By lifting the flap (facing page), the reader was presented with the company's message, that yes, there is a better way to move cargo - the Chicago Tunnel Company.

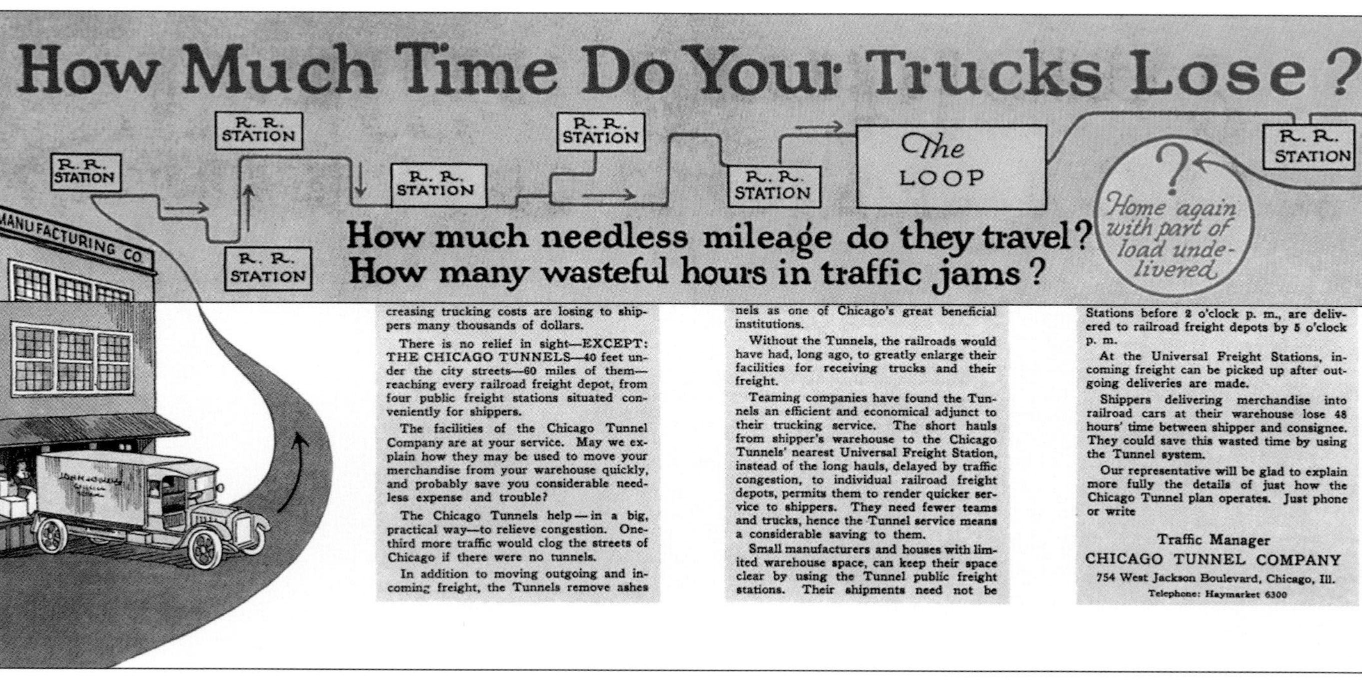

Can We Solve Your Traffic Problem?

Your shipments are delivered in a few minutes to the nearest

UNIVERSAL FREIGHT STATION

R. R. STATION

R. R. STATION

R. R. STATION

R. R. STATION

R. R. STATION

R. R. STATION

R. R. STATION

All shipments received before 2 o'clock are at respective railroad freight depots before 5 P.M.

MANUFACTURING CO.

EVERY shipper in Chicago—large or small—is confronted by a shipping problem that is becoming more serious daily.

Congestion of traffic on Chicago streets and at railroad freight depots is becoming intolerable. Consequent delays and increasing trucking costs are losing to shippers many thousands of dollars.

There is no relief in sight—EXCEPT: THE CHICAGO TUNNELS—40 feet under the city streets—60 miles of them—reaching every railroad freight depot, from four public freight stations situated conveniently for shippers.

The facilities of the Chicago Tunnel Company are at your service. May we explain how they may be used to move your merchandise from your warehouse quickly, and probably save you considerable needless expense and trouble?

The Chicago Tunnels help—in a big, practical way—to relieve congestion. One-third more traffic would clog the streets of Chicago if there were no tunnels.

In addition to moving outgoing and incoming freight, the Tunnels remove ashes from buildings and make coal deliveries to them. Through them have been moved the thousands of tons of dirt and refuse that made the land on which the new Field Museum stands.

Many public spirited citizens have expressed their regard for the Chicago Tunnels as one of Chicago's great beneficial institutions.

Without the Tunnels, the railroads would have had, long ago, to greatly enlarge their facilities for receiving trucks and their freight.

Teaming companies have found the Tunnels an efficient and economical adjunct to their trucking service. The short hauls from shipper's warehouse to the Chicago Tunnels' nearest Universal Freight Station, instead of the long hauls, delayed by traffic congestion, to individual railroad freight depots, permits them to render quicker service to shippers. They need fewer teams and trucks, hence the Tunnel service means a considerable saving to them.

Small manufacturers and houses with limited warehouse space, can keep their space clear by using the Tunnel public freight stations. Their shipments need not be sorted by destinations or railroads and need not accumulate. When a shipment is ready, it is hauled immediately to the nearest Universal Freight Station. There it is loaded into Tunnel cars and routed to the proper railroad freight stations.

Shipments received at Universal Freight Stations before 2 o'clock p. m., are delivered to railroad freight depots by 5 o'clock p. m.

At the Universal Freight Stations, incoming freight can be picked up after outgoing deliveries are made.

Shippers delivering merchandise into railroad cars at their warehouse lose 48 hours' time between shipper and consignee. They could save this wasted time by using the Tunnel system.

Our representative will be glad to explain more fully the details of just how the Chicago Tunnel plan operates. Just phone or write

Traffic Manager
CHICAGO TUNNEL COMPANY
754 West Jackson Boulevard, Chicago, Ill.
Telephone: Haymarket 6300

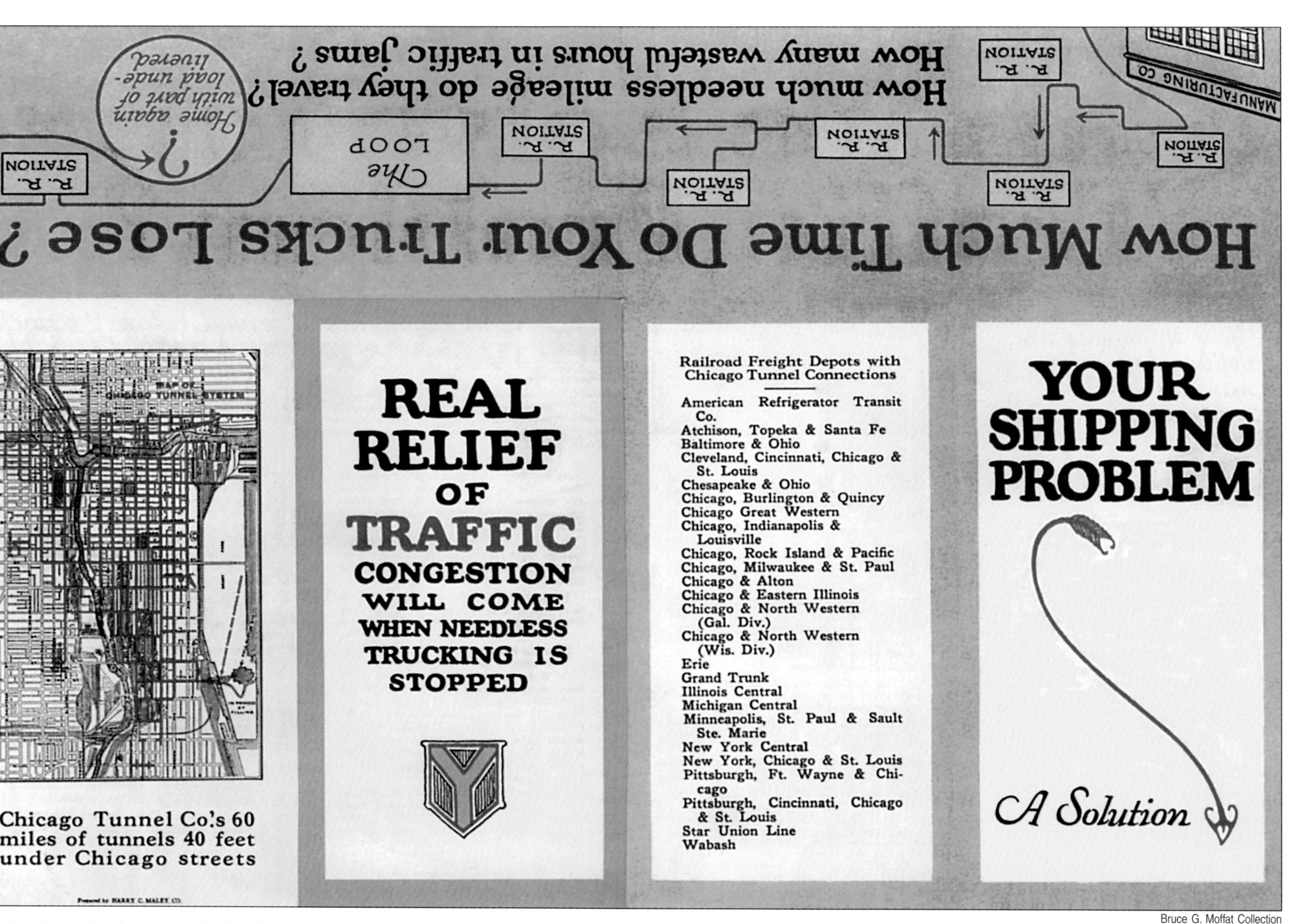

Bruce G. Moffat Collection

Issued in the mid to late 1920's this brochure included a listing of major factories and warehouses served by the little trains (above) as well as a system map (facing page).

Can We Solve Your Traffic Problem?

Your shipments are delivered in a few minutes to the nearest

MANUFACTURING CO.

UNIVERSAL FREIGHT STATION

R. R. STATION

All shipments received before 2 o'clock are at respective railroad freight depots before 5 P.M.

EVERY shipper in Chicago—large or small—is confronted by a shipping problem that is becoming more serious daily.

Congestion of traffic on Chicago streets and at railroad freight depots is becoming intolerable. Consequent delays and increasing trucking costs are losing to shippers many thousands of dollars.

There is no relief in sight—EXCEPT: THE CHICAGO TUNNELS—40 feet under the city streets—60 miles of them—reaching every railroad freight depot, from four public freight stations situated conveniently for shippers.

The facilities of the Chicago Tunnel Company are at your service. May we explain how they may be used to move your merchandise from your warehouse quickly, and probably save you considerable needless expense and trouble?

The Chicago Tunnels help—in a big, practical way—to relieve congestion. One-third more traffic would clog the streets of Chicago if there were no tunnels.

In addition to moving outgoing and incoming freight, the Tunnels remove ashes from buildings and make coal deliveries to them. Through them have been moved the thousands of tons of dirt and refuse that made the land on which the new Field Museum stands.

Many public spirited citizens have expressed their regard for the Chicago Tunnels as one of Chicago's great beneficial institutions.

Without the Tunnels, the railroads would have had, long ago, to greatly enlarge their facilities for receiving trucks and their freight.

Teaming companies have found the Tunnels an efficient and economical adjunct to their trucking service. The short hauls from shipper's warehouse to the Chicago Tunnels' nearest Universal Freight Station, instead of the long hauls, delayed by traffic congestion, to individual railroad freight depots, permits them to render quicker service to shippers. They need fewer teams and trucks, hence the Tunnel service means a considerable saving to them.

Small manufacturers and houses with limited warehouse space, can keep their space clear by using the Tunnel public freight stations. Their shipments need not be sorted by destinations or railroads and need not accumulate. When a shipment is ready, it is hauled immediately to the nearest Universal Freight Station. There it is loaded into Tunnel cars and routed to the proper railroad freight stations.

Shipments received at Universal Freight Stations before 2 o'clock p. m., are delivered to railroad freight depots by 5 o'clock p. m.

At the Universal Freight Stations, incoming freight can be picked up after outgoing deliveries are made.

Shippers delivering merchandise into railroad cars at their warehouse lose 48 hours' time between shipper and consignee. They could save this wasted time by using the Tunnel system.

Our representative will be glad to explain more fully the details of just how the Chicago Tunnel plan operates. Just phone or write

Traffic Manager
CHICAGO TUNNEL COMPANY
754 West Jackson Boulevard, Chicago, Ill.
Telephone: Haymarket 6300

Bruce G. Moffat Collection

Issued in 1915, Lifting the Lid In The Loop *explained how different commodities were handled and interchanged with other carriers.*

Kogan Collection

"Why don't you come down and see me some time?" Sex symbol Mae West poses on an inspection car opposite the Palmer House siding on Wabash Avenue in the late 1920's. Note that "Palmer" has been misspelled as "Plamer" on the wall.

Going Hollywood

Aside from a few newsreel appearances, the freight tunnel system was woven into the story line of apparently only one major Hollywood production during its half-century of operation: "Union Station," a 1950 Paramount release starring William Holden, Nancy Olson and Barry Fitzgerald. In this story, William Holden is a railroad police lieutenant on the trail of a kidnapping ring that had abducted the blind daughter of a wealthy businessman. Set in an unnamed city, most of the action actually takes place in the Los Angeles Union Passenger Terminal. In one long segment, however, the heroine has been temporarily sequestered in the "Municipal Tunnel" while the ringleader attempts to collect the ransom.

Built on a Hollywood sound stage, the tunnel set had some of the trappings of the genuine article including three-way switches, overhead wire and a trolley-style locomotive complete with merchandise car. Although generally convincing, the "tunnel's" much larger profile and out-of-scale appearance give it away.

Thirty years later, the tunnels served as the inspiration for a brief sequence in Universal's 1981

Bruce G. Moffat

The same location in 1980.

Sveriges Television Collection

State and Madison – "The World's Busiest Corner" as it looked in December 1926 from forty feet above. The appearance was quite different at forty feet below (right).

release, "The Blues Brothers," starring former Second City and "Saturday Night Live" regulars Dan Aykroyd and the late John Belushi in an expansion of their successful SNL routine. Dressed in their funereal garb of black suits, dark glasses and pork pie hats, the "Blues Brothers" attempt to save the Chicago orphanage in which they had grown up from being closed for failure to pay back taxes. They set out to raise the necessary $5,000 by holding a benefit concert while remaining one step ahead of the police.

One of the film's sequences has the brothers escaping the long arm of the law after the concert through the use of an emergency exit from the concert hall (actually Chicago's South Shore Cultural Center) that conveniently opens into the tunnel system.

Unsuccessful in their efforts to obtain permission to shoot the sequence in an actual tunnel segment, a section of the Randolph tunnel was carefully reproduced on a Hollywood sound stage. Complete with track, the screen version of the tunnel has only two errors that betray it as a replica. The entranceway used by the brothers is of rough-hewn stone (unlike the poured concrete of the real tunnel), and overhead lighting replaces the side-mounted variety of the

Bruce G. Moffat Collection

bona-fide bore. Had the tunnel sequence been filmed "on location," it would have joined the many other scenes that were filmed in the Windy City (numerous elevated trains and careening police cars found their way into the film). Perhaps the film's producers were unaware of the tunnel segment controlled by the Field Museum. Complete with a surviving (although derelict) locomotive its larger diameter bore would have greatly minimized the logistical problems of filming in cramped quarters.

The 1990's have seen the tunnels appear briefly in a few television programs and well as in a number of documentaries on transportation, tunnels and urban development.

The Golden Age of Radio

During the 1930s and 40s, network radio was enjoying its "Golden Age," offering listeners an incredible array of dramatic and non-dramatic fare of a quality seldom equaled in present-day broadcasting.

One of the more renowned programs was a CBS-aired radio series called "Lights Out," which offered suspenseful tales of the mysterious and bizarre. The chief writer for the series (which began in Chicago in 1934) was Arch Oboler, who as a youngster was treated to a tour of the sub-sub-basement of a large downtown department store (most probably Marshall Field & Company or Carson Pirie Scott & Company). The tour included a glimpse into the freight tunnel which served the building, and from this early exposure came the inspiration for the August 24, 1943 entry in the "Lights Out" series, entitled "Sub-Basement."

The story opens with a very bitter Arnie taking his wife Irma on an after-hours tour of the sub-basements of a "huge downtown department store" where he works as a supervisor. Arnie is jealous of his wife of ten years, and believing that she is too good for him, plans a murder-suicide in the freight tunnel.

As the story unfolds, he leads her through the building's various sub-basements to the tunnel entrance where he describes the miniature trains, which deliver merchandise to all the department stores in the "business section." The trains are powered by small battery locomotives (in a bit of artistic license) like those used in mines.

Once in the tunnel, he tells her "this is as far as we're going," and prepares to kill her. He is stopped, however, by her claim of having "heard something."

As they begin to walk back to the store, they stumble upon Tom, a security guard, dead – his throat savagely ripped out. Frightened, they run back to the store's tunnel entrance, only to find it locked and a nearby emergency exit mysteriously blocked off.

Now very terrified, they begin walking through the tunnel toward another department store where, Arnie explains, they can find another exit. Before they can reach that point, however, they find their path blocked by a prehistoric lizard-like "dinosaur" – which apparently entered the system through a hole where blasting for a tunnel extension had recently taken place. They manage to evade the creature only to have Arnie break his leg. He then tells Irma of his plan to kill her and urges her to escape before the creature

Bruce G. Moffat Collection

The bogus "State & Madison" under the office building. Workmen are supposedly listening to a safety speech by P. E. Crowley, president of the New York Central Railroad. The date is May 5, 1929.

(which they have determined to be blind) finds them.

As the story concludes, Irma tearfully (and reluctantly) flees to safety while Arnie is left to meet the same fate as Tom. The listener is left to assume that the creature later disappears or is killed and that the tunnels are once again made safe for less-than-carload freight.

The World's Busiest Corner

From the early years until well past mid-century, the intersection of State and Madison Streets in downtown Chicago was touted by civic boosters as the "world's busiest corner." Whether or not it actually was is irrelevant, because the claim achieved its purpose of luring tourists to the Windy City to visit the ennobled intersection.

The reputation of this famous crossing led to numerous requests by those fortunate enough to be granted an escorted tour of the tunnels to have their photograph taken next to a sign identifying this point underground. Honoring such requests generally meant that a locomotive, inspection car, photographer and motorman would have to be more or less instantly available, with the resultant disruption of regular tunnel freight-hauling duties.

The problem of honoring these entreaties without having to transport guests from the offices on Jackson Boulevard over to State and Madison was solved with typical Tunnel Company ingenuity. They simply created another "State and Madison" intersection near the office. This was accomplished by applying whitewash to the walls where a spur track left one of the main bores. A stenciled "State & Madison" gave the location an air of authenticity and was used as the background for many publicity photos.

Although it lacked the grand union of its legitimate counterpart, the application of a coat of whitewash to walls and a stenciled street sign were all that were needed to lend an air of legitimacy to the scene. This same location was used on occasion by newspaper photographers who needed a tunnel shot to illustrate a story. By World War II, however, the company's passion for publicity had palled and visitors were few. The use of the fake intersection for photographs had been discontinued and the painted markings were apparently removed.

8 False Hopes and Failure

Vitality and aggressiveness were two terms which could safely be said did not apply to the tunnel's management team in the later years. From the early 1930s on, the company strove to maintain a low public profile while watching its business slip away. The appearance in 1954 of a single individual, however, promised to change all that.

William Henning Rubin was a newcomer to the "railway under the Loop," but he was not a stranger to Chicago. Among other enterprises, he ran the famed Morrison Hotel, a downtown landmark (since razed) that once billed itself as the world's tallest hotel (41 stories). By 1954, Rubin had managed to purchase a 30% interest in the Chicago Tunnel Terminal Corporation. This was sufficient for him to be named to the chairmanships of the Chicago Tunnel Company and the Chicago Tunnel Terminal Company.

One of his first orders of business was to cut operating expenses. To assist in this effort, he hired a New York-based management consulting firm to study the system and make recommendations. He also actively solicited new business, something that had not been done in years. Although his efforts to stimulate new traffic were largely unsuccessful, he was able to breathe some life into a long-standing proposal to again operate mail trains.

CB&Q Railroad

Looking into the Burlington Building's connection from Clinton in 1946. The number on the wall identified the siding's customer.

Bruce G. Moffat Collection

(Left to right) Earl H. Chambers, Assistant General Superintendent, Postal Transportation Service for the Post Office Department; Edward Freeman, Assistant Director, Air Service, Central Region U.S. Post Office Department; Joel Rubin; and William H. Rubin, Chairman of the tunnel company pose in the tunnel beneath the city's main post office for the photographer.

Moving Mail

The idea to revive mail operations apparently originated, not with the Chicago Tunnel Company, but with the United States Post Office. On December 6, 1951, Chicago Postmaster John Haderlein disclosed that engineering feasibility studies were being conducted to determine whether the use of the tunnels could be justified. Haderlein estimated that, by using the tunnels, the Post Office Department could save two or three million dollars per year in operating costs, remove 300 to 400 trucks from downtown streets, and speed up mail delivery.

Over short distances it was proposed to use conveyors. Automated trains would be used tohandle the longer hauls. Initially, the main post office and six railroad terminals were to be served, with larger downtown buildings being connected at a later date. To meet postal requirements, the bores would also have to be widened by 1-1/2 feet to allow for supervision of the mails.

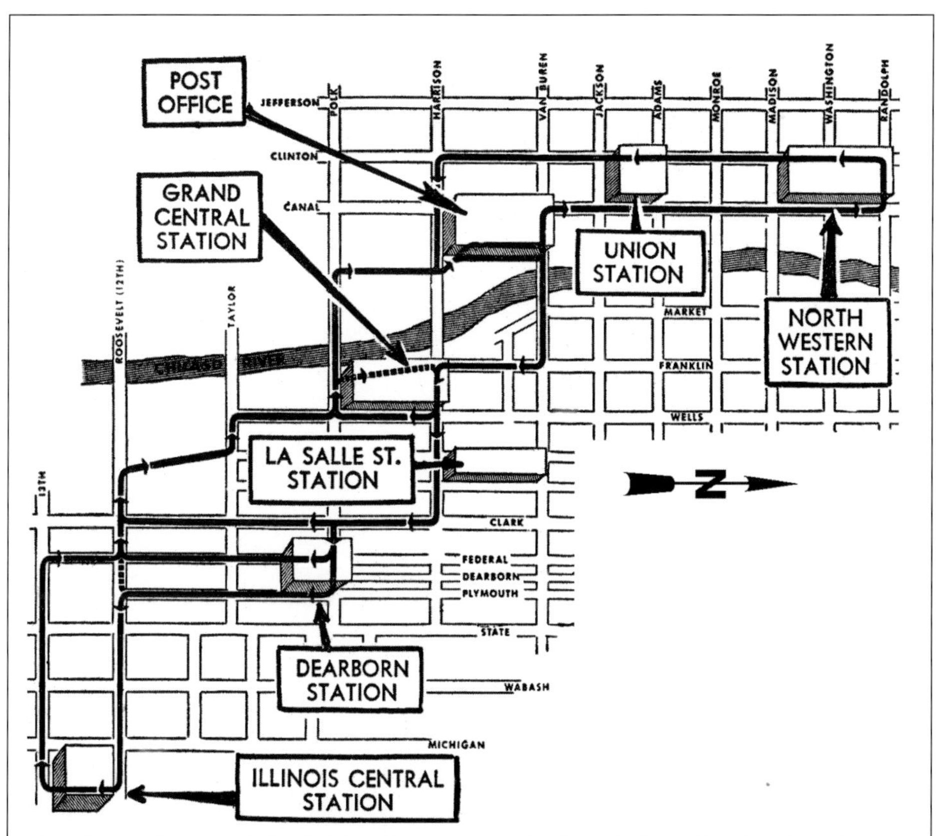

Chicago Daily News

A 1953 map of the proposed mail train system. Only Union Station still boasts intercity rail service and handles mail.

Bruce G. Moffat Collection

On August 27, 1954, William Rubin (second from left) joins postal officials at Chicago's new main post office on Canal Street to witness the loading of a cart loaded with dummy mail onto a waiting flat car for delivery to the B&O Railroad's freight house. Although the test was a success the company was unable to obtain a mail contract.

Bruce G. Moffat Collection

Chicago Tunnel Company president George A. Kirk stands next to one of the two cars used in the experimental mail run from the main post office to the Monon's freight house on July 14, 1954.

Haderlein's bosses in Washington were not enthusiastic about this concept and refused to authorize the use of the tunnels for handling mail. In 1953, the Merchandise Delivery Company attempted to interest postal officials in a proposal to allow the company to provide mail delivery service through the tunnels under contract. This entreaty was accompanied by letters of support from the State Street Council, the Chicago Association of Commerce and Industry and the Illinois Small Businessmen's Association. These business groups agreed that the tunnels could help relieve traffic congestion, reduce transportation costs and speed up mail delivery. Postal officials in Washington remained unconvinced, however.

On October 14, the city council entered the fray when alderman Roy Olin of the 8th Ward, and chairman of the Committee on Railway Terminals, proposed that his committee study the possibilities. His resolution noted that, in addition to eliminating 3,000 daily truck movements from downtown streets, the use of the tunnels would increase city revenues from tunnel operations. By 1950, the city was receiving only the $10,000 per year minimum required under the franchise because of declining traffic.

In July 1954, Rubin finally prevailed on postal officials to at least authorize a single test trip to determine if the concept was feasible. Rubin of course had every confidence that it was. On July 14, with postal

Bruce G. Moffat Collection

George A. Kirk stands at the telephone dispatching console at the company's new headquarters in leased space at 340 W. Harrison Street on February 28, 1950.

officials and other individuals looking on, sixty sacks of mail weighing six tons were loaded into two tunnel cars (which had been specially painted for the occasion) at the south end of the city's main post office complex at Harrison and Canal Streets. The two cars were then towed to the Monon freight house located just west of Dearborn Station on Federal Street. The Monon facility was used because it was vacant and would not cause interference with regular mail movements.

Although Rubin termed the trial a success, postal officials remained unconvinced. The fact that the origin and destination points were only a half-mile apart and could be traveled by truck in a fairly short period of time did not help matters.

Not one to give up easily, Rubin arranged for a second trial run which took place on August 27. In this demonstration a forklift was used to place a four-wheel loaded mail cart on a flat car. The new destination was even closer to the main post office: the Baltimore & Ohio Railroad's freight house on Wells Street south of Polk Street. Following this demonstration, further attempts to win over postal officials were abandoned, and hopes for revitalizing the system dimmed appreciably.

Jeffrey Division, Dresser Industries

Class 30 Jeffrey six-ton locomotive #406 looks much the worse for wear (and poor maintenance) in this 1957 work train photo after a half century of tunnel service.

Receivership

The sudden change in corporate personality wrought by Rubin was not without its casualties. On July 27, 1954, the entire executive staff of the two subsidiary companies quit in protest over what they alleged to be impractical operating methods that he was implementing.

Among the resignations was that of George A. Kirk, president of both companies and a 47-year veteran of the system. Another was vice president and treasurer Rudolph Tenicki, who had been with the companies for 31 years. Rubin subsequently assumed the presidency vacated by Kirk, while Tenicki was eventually persuaded to return to his post. Rubin was, however, unable to halt the system's economic slide, resulting in his replacement by Tenicki in 1955. That year also saw the general offices being moved again. The new address was the public station at 566 North Kingsbury Street.

Unable to meet their financial obligations, both the Chicago Tunnel Company and the Chicago Tunnel Terminal Company applied for reorganization under Section 77 of the Bankruptcy Act on May 4, 1956. The case was assigned to U.S. District Court Judge Julius Hoffman.

On July 11, Judge Hoffman appointed George W. Lennon as receiver for the two companies. Interstate Commerce Commission concurrence was given two weeks later. ICC approval was necessary because of the interstate nature of the company's business (a carrier need not actually operate across state lines to be under ICC jurisdiction.)

In September 1956, all LCL operations were abandoned, leaving only ash and some trash removal to the little trains. The public stations were closed and the offices were moved once again – this time to 2600 South Normal Avenue, home of the Norwalk Truck Line. This move left the Chicago Tunnel Company without direct access from its offices to the tunnels for the first time. With the public stations now closed, the Chicago Tunnel Terminal Company's few remaining

Bruce G. Moffat Collection

Time was running out for the "railroad under the Loop" when this photograph was made in 1955. Note the close proximity of the energized overhead wire to the motorman's head.

Ted Koston

Above: *The company maintained its own small security staff at the public stations to deter theft.*

employees were left with only the Grand Avenue disposal station to operate.

On April 1, 1958, the offices were moved yet again; this time to 445 North Lake Shore Drive, home of the North Pier Terminal Company's warehouse. At least this building had tunnel access. In moving from the Normal Avenue location, the companies left behind an unpaid office rental bill of $33,600.

On June 19, 1958, a last-ditch effort to save the system was made by an unaffiliated organization, the Chicago Tunnel Users Committee. The Committee represented about 20 remaining tunnel customers who hoped to purchase the system in order to continue the ash removal function. Members included the owners or managers of a number of major Loop buildings, including Marshall Field & Company, the Chicago Tribune, the Merchandise Mart and the Morrison Hotel. The Committee made an offer of $225,000 to purchase the system and reorganize it into a single company.

Negotiations dragged on into early September, when the Interstate Commerce Commission decided that their offer was insufficient to cover the ever-increasing deficit and legal expenses.

Meanwhile, in an effort to satisfy creditors, a bankruptcy sale was held, resulting in the transfer of control of the two companies to the Rutland Transit Company, a Vermont-based trucking company, on August 19. Any reorganization plan would now have to be approved by Rutland, as well as the court and the ICC.

By March 1959, the financial situation had turned from precarious to hopeless. On March 2, the Chicago& North Western Railway cut off all credit to the Chicago Tunnel Terminal Company following its fail-

Marshall Fields

Locomotive 533 and "crew" maneuvering an ash car in the second subbasement of Marshall Field's. Note the pail (also numbered 533) which carried sand that the motorman would distribute on slippery rail by means of an old coffee can.

ure to pay nearly $15,000 owed for hauling ashes away from the Grand Avenue station and for 14 months back rent on the property. In addition, the company handling the actual disposal of the cinders refused to accept further shipments until it received an equal amount. Although it now had no place to dump ash, the company continued to load cars until March 6. Some of these loaded cars could be found languishing in the tunnel network decades later.

Unable to raise the necessary additional capital from its members to purchase the operation, the Users Committee withdrew its offer on March 5. With all hope of continuing operations gone, the Tunnel Company discontinued ash-loading operations the next day. Some non-revenue train movements continued through March 9, at which time Judge Hoffman ordered the trustee to cease all but pumping operations.

Applications for abandonment on behalf of both the Chicago Tunnel Company and the Chicago Tunnel Terminal Company were filed with the Interstate Commerce Commission on March 16, and with the Illinois Commerce Commission on March 17.

In response to a questionnaire from the Interstate Commerce Commission, the Chicago Tunnel Company gave the following tersely-worded justification for requesting abandonment:

> The applicant's business of hauling cinders and debris declined as a result of loss of business to trucks, which, unhampered by rates fixed by the Interstate Commerce Commission, removed the cinders and debris for less than the rates fixed for the Chicago Tunnel Company, and as a result of loss of business, revenue declined which in turn resulted in the unavoidable depreciation of equipment because of lack of funds for necessary repairs which only resulted in inefficiency and further loss of revenue.

"Unavoidable depreciation" was probably an understatement. A study prepared in early 1959 (when the system was still operating), by George W. Barton & Associates, listed only four locomotives out of a fleet of 83 that could be considered usable; and of these, only two or three were actually operational. Most of the ash cars on the property were in need of reconditioning, while only 8 of 71 pumps operating in 1928 were still functioning. Also, by this time only 34 ash

Bruce G. Moffat Collection

The company's elevators lacked doors at tunnel level as graphically illustrated in this view from February 23, 1937.

connections were still in use, while none of the 60-odd railroad freight house spurs that were operation at one time were still active.

Testimony before the Illinois Commerce Commission on May 11, 1959, cast further light on the rundown condition of the property. When ash operations were halted, 98 loaded cars were left stranded throughout the system and the operable locomotive fleet had dwindled to a single unit. Pump failures had caused the Randolph, Franklin and Orleans river crossings to flood completely, effectively isolating portions of the system. Operating losses were averaging $9,000 per month prior to the shutdown. The combined staff of the two companies numbered just six: two watchmen, two pump repairmen, a secretary, and president Tenicki.

No objections to the abandonment petitions were heard at hearings held by both the Illinois or Interstate Commerce Commissions, clearing the way for the Interstate Commerce Commission to approve the petition on June 10, with the Illinois Commerce Commission following suit on July 13.

One of the world's most unusual railroads was now little more than a historical footnote in Chicago's evolution.

Glenn M. Andersen

In April 1959, during the very last days of the tunnel company's life, a local rail enthusiast was given rare access to the one of the sidings beneath the company's North Pier Terminal headquarters and recorded these views.

Dismantling the System

With rail operations now history, Lennon turned his attention toward the disposition of all saleable equipment. Time was of the essence as the lack of operable pumps was already starting to create havoc according to one newspaper report:

> **Reports Peril of Water In Tunnel**
>
> Water rising in Chicago's defunct underground railway is threatening downtown buildings, a trustee in bankruptcy said Thursday in U.S. District Court.
>
> George W. Lennon, who has been managing the railroad's operating companies, the Chicago Tunnel Co. and the Chicago Tunnel Terminal Co., for the last three years, said:
>
> "The Merchandise Mart reported it was within one foot of trouble." And he added there could be a "slight danger" to the roundhouse at Union Station which is located underground.
>
> However, Walter A. Stahl, operations manager for the Mart, said later that the water would have to rise an additional five feet before he would consider it dangerous.
>
> After hearing Lennon's report, Judge Julius H. Miner granted an emergency plea to file an equity receivership suit so the railway equipment can be sold.
>
> The companies, which halted half a century's operations by court order in March, leased the tunnel space from the city. Pumps are being used in an effort to control the water level in the tunnel.
>
> The city is expected to retake possession of the 47 miles of tunnel once legal complications are cleared. The railway was used to haul cinders, trash and parcels during solvent times.

Put up for auction were the locomotives, rolling stock and 215 tons of copper trolley wire and lead-covered feeder cable. Also included were the two stiff-legged derricks used at the Grand Avenue disposal station, 25 elevators and miscellaneous machinery and scrap. Not for sale, however, was the track, which was embedded in the concrete tunnel floor and was deemed too costly to remove.

On October 29, 1959, about 25 interested bidders converged on North Pier Terminal for the auction. Sample cars were positioned on the sidings under the building for inspection.

The successful bid was entered by Sam Fishman, a North Side scrap dealer representing eight of his fellow junkmen. His offer of $64,000 was $16,000 below Trustee Lennon's estimate, which was based on a scrap value of $140,000 less $60,000 in removal costs.

On October 29, Judge Hoffman approved the sale, permitting Fishman and his associates to begin dismantling the system. The overhead wire was cut down in short lengths, leaving the hardware fittings still attached to the ceiling. Also removed were those cars that were readily accessible and still capable of movement. Left behind were a number of ash and merchandise cars. Some were abandoned where they had derailed, while others were seemingly just forgotten at other points. Most remain in the tunnels to this day. Also left behind were two locomotives, trolley wire fixtures and several pumps.

All that remained to be done was some final bookkeeping and filing a petition with the state to dissolve the two corporations. The companies were formally recorded as dissolved by the State of Illinois on December 27, 1960.

9 Affiliated Enterprises

Further complicating an already complex family tree were several "side businesses" that were a natural outgrowth of the tunnel's normal operations. For a time, these businesses helped keep the system solvent. Unfortunately they were unable to maintain their profitability in Chicago's rapidly changing business climate.

Chicago Tunnel Transport

An increase in activity by competing trucking companies, coupled with the need to increase freight revenues (and reach potential customers not served by the existing tunnel system), led to the incorporation of the Chicago Tunnel Transport Company. Incorporated on May 16, 1927, as a wholly owned subsidiary of the Chicago Tunnel Terminal Corporation, its function was to provide pickup and delivery service between shippers and the public stations of the Chicago Warehouse & Terminal Company.

Like its sister companies Chicago Tunnel Transport was not without its share of controversy. It was alleged that the company was formed simply to divert freight from the tunnel trains in order to reduce the amount of compensation owed the city. (The city collected a franchise fee based on the amount of freight moved through the tunnels but not on freight moved by truck.)

Initially, the trucking company was quite successful. By 1946, it employed 74 men and rostered a fleet of 66 tractor-trailer combinations, three one-ton trucks and an assortment of trailers. Unfortunately,

Chicago Transit Authority

Displaying a sign reading "Pick up and Delivery – Tunnel and Rail," a Chicago Tunnel Transport truck threads it way past the railroad freight houses on Clark south of Polk.

The early 1950's found Chicago Tunnel Transport still shuttling freight around Chicago. Tractor #15 is seen here at Public Station #2 on Kingsbury positioning a trailer at the loading dock.

Bruce G. Moffat Collection

the post-war years brought a general decline in traffic as businesses moved away from the Loop and the railroads lost their monopoly on LCL traffic. Red ink soon made its appearance on the company's books resulting in all operations being discontinued in September 1956, idling 43 truck tractors and 310 trailers. The company was formally dissolved on December 18, 1959.

Chicago Steam Corporation

In 1933, the last enterprise to be affiliated with the tunnel system was incorporated to transport a commodity that could not be handled like other cargo – steam.

The Chicago Steam Corporation was wholly owned by the Chicago Tunnel Terminal Corporation and was created to supply steam heat to buildings in the area bounded by the Chicago River on the west and north, Polk Street and Michigan Avenue. Except for a few locations where it was proposed to install the steam pipes below the tunnel floor, all ducting was to be installed at track level, making further operation of trains in certain tunnels impossible.

As lessee of the now city-owned bores, Chicago Tunnel applied on two occasions for a loan from the Reconstruction Finance Corporation to fund the project. The RFC rejected the applications, but it was decided to proceed with the project anyhow, albeit in a greatly curtailed form. The steam heating project would be limited to the area north of Jackson and east of the river. Accordingly, an agreement was reached with the Chicago Union Station Company to purchase steam generated by its powerhouse at Taylor Street and the river for resale.

Most of the pipes were suspended from the ceiling, allowing for continued train operations. In the Jackson Boulevard tunnel, however, a large diameter pipe was laid directly on pads mounted on the tunnel floor, precluding train operations in that tunnel between Canal and Clark Street, and on Market Street between Jackson Boulevard and Adams Street. Installed during 1934, the pipe also cut off rail service in several intersecting tunnels. A "bridge" arrangement was installed at Jackson and Sherman so that trains could cross the steam line enroute to or from the southern portion of the system. Ultimately, Chicago Steam served just 12 buildings, generally clustered in the vicinity of Jackson and Franklin.

Once steam service began, it did not take long for local coal suppliers to organize their opposition. In 1935, a complaint was filed with the Illinois Commerce Commission against Chicago Steam and the Chicago Union Station Company, asking that the Commission order the service halted because the companies failed to obtain a certificate of convenience and necessity from the regulatory agency. The city also intervened in the suit, taking the position that the Tunnel Company was not authorized under its franchise to transmit steam through the tunnels without municipal permission. Fiscal reason soon prevailed, however, and the city agreed to accept a payment of 3 cents per thousand pounds of steam carried.

About 1939, for reasons that are now unclear, the Pennsylvania Railroad, as part owner of the Chicago Union Station Company, ordered Chicago Union Station to discontinue the sale of steam to Chicago Steam forcing the latter firm to suspend operations.

This action left four buildings occupied by war agencies and other industries without any alternative heating sources. Since wartime conditions meant that new boilers for these buildings were unavailable, the Chicago & North Western Railway, owner of one of the buildings, contracted with CUS to supply steam which the Chicago Tunnel Company then transported via the existing steam pipes. This arrangement lasted until 1948.

Steam service returned to the tunnels in a limited fashion in the early 1970's with several small installations. One connected the Chicago Public Library's old main library (now the Chicago Cutural

Bruce G. Moffat

A modern day view showing the steam lines (attached to wall) that run for a short distance through the Wabash tunnel south of Adams to link two buildings occupied by DePaul University. The overhead conduits house fiber optic communication lines.

Center) at Washington and Michigan with the boilers of the nearby Pittsfield Building. Another connected the Burlington Building at 547 W. Jackson with Union Station where a connection was made with the steam supply pipe from their powerhouse. Finally, a small diameter pipe was installed through the tunnel to connect two buildings occupied by DePaul University at Jackson and Wabash.

City News

In addition to their other functions, for many years the tunnels housed the pneumatic tube system operated by the City News Bureau. City News was a news-gathering cooperative owned by local newspapers and radio and television stations and was located just a short distance from City Hall. Reporters returning from City Hall would prepare their reports and submit them to an editor for review. The reports were then placed in message cylinders and dropped into the tube. Propelled by compressed air, the cylinders were whisked through the tunnels to the editorial rooms of the member news organizations.

By the 1970s, City News had switched to electronic data transmission and the tube system was removed. The Bureau itself was dissolved in 1999.

Air Conditioning

Before the advent of modern air-conditioning systems, the cool tunnel air (generally about 55 degrees) was sold to adjacent buildings. Movie theaters were the principal users of this offshoot of tunnel services, keeping their auditoriums cool at low cost during the typically hot Chicago summers.

At one time, more than 20 buildings availed themselves of the tunnel's cool air. The Tunnel Company apparently sold this primitive form of air conditioning on a flat rate basis to interested buildings. As large-scale mechanized air conditioning systems became practical, the use of tunnel air was gradually abandoned.

SECTION
IV
The Tunnels Today

10 Exploring the Railroad "Forty Feet Below"

Unlike most former railroads where scant physical evidence of their existence remains four decades after abandonment, the Chicago Tunnel Company's physical plant has remained largely intact. Unfortunately, unlike other "ghost" railroads, it is nearly impossible for the casual observer to take note of those remains because of their vertical isolation.

With the dissolution of the Chicago Tunnel Company in 1960, the city was now responsible for maintaining what it had claimed ownership of decades earlier. A handful of relatively small pumps were installed at a few of the river crossings near City Hall to maintain access to areas west and north of the river for occasional inspections. Although nearly all of the track remained in place, the presence of dirt and other debris, inoperable switches, and lack of electricity, precluded its use to transport the inspection teams. This meant that all inspections had to be made on foot.

For many years the city's inspection and maintenance team consisted of former tunnel employees Julian Waisnor and Ignatius DeCicco. Their supervisor was former president Rudolph Tenicki. In 1963, a Chicago *Tribune* reporter joined Waisnor and DeCicco on a short walking tour and wrote what was probably the first of many "filler" news articles about the abandoned railroad under the Loop. Waisnor, 53, reflected on his joining the company in 1931: "That was depression time, I was lucky to get any kind of job. They paid me 65 cents an hour for a 48-hour week. In those days that was good money." The tunnel veteran noted that due to his long association with the company, he did not need a map or guide markers to help him get around: "Even when the flashlight has gone out, I still get around. It is like being in your own home in the dark. You just seem to know where the doors are and how to avoid the furniture. I have never known anyone to get lost down here. Once in awhile, there is a stray cat, but they do not stay long."

Despite the inspections, it would be inaccurate to say that the tunnels were kept even reasonably dry. Some sections had been rendered impassable due to high water years earlier, and there was little incentive to justify maintenance of the old positive displacement pumps to reach the system's extremities. Locations isolated by localized high water included the Grand Avenue undercrossing of the State Street subway, the Jackson Boulevard drift under the Chicago River and some of the tunnels south of Harrison Street.

Once in a while the seepage tended to cause problems. One of the more noteworthy occurrences happened in 1963 when the Van Buren Street river crossing filled with water. The water eventually seeped

Bruce G. Moffat

Left: *In 1971, former Tunnel Company employees Julian Waisnor and Ignatius DeCicco were still walking the tunnels as inspectors for the Chicago Department of Public Works. Wisnor hired on with the Tunnel Company in 1931. DeCicco started two years later.*

Chicago Tribune

Right: *Standing about four feet tall, this abandoned pump was one of dozens that were used to keep water seepage under control.*

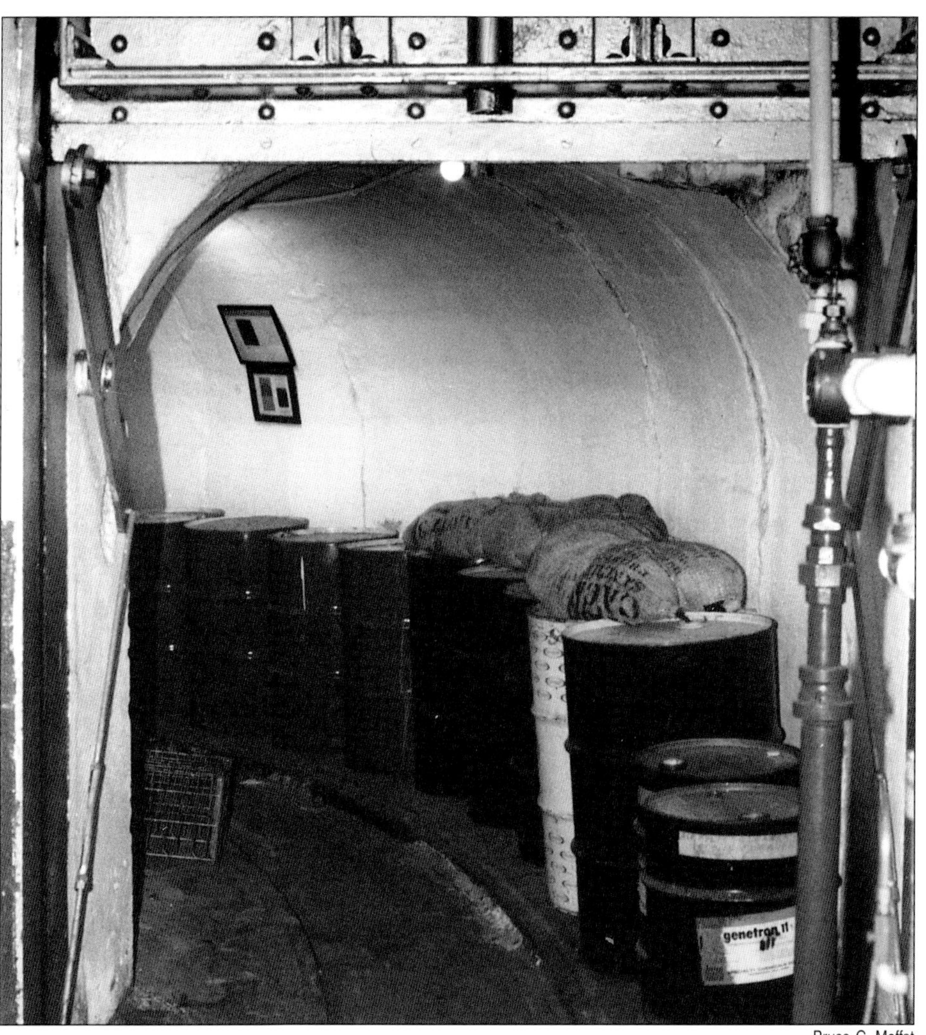

Bruce G. Moffat

Right and below:
The Chicago Tribune Company converted a segment of the tunnel that served their headquarters building on Michigan Avenue into storage space following the system's 1959 abandonment.

Bruce G. Moffat

into an underground cable vault belonging to the Illinois Bell Telephone Company, knocking out service to some 1,500 telephones before the water could be pumped out and the cables dried. Clearly, maintenance was kept to a minimum.

Some Possible Uses

Over the years, numerous suggestions have been made to city officials for coverting the system to other uses. Ranging from the commonplace through the unusual to the downright bizarre, these proposals included converting the tunnels into an underground mushroom farm, bomb shelters (in the event of a nuclear attack), a steam-heat conduit network, and even a downtown people-mover.

One of the principal proponents of the people-mover concept was Larry S. Bell, a professor of industrial design with the University of Illinois in Urbana. Professor Bell supervised two studies for the city during 1972. His plan called for the operation of both single units and trains with a capacity of up to 11,000 riders per hour on any one line. Top speed would have been about 20 mph. Little interest was shown by the city in the use of the tunnels for this proposal and the idea was quietly shelved.

Bruce G. Moffat

When the Milwaukee Road freight house at Carroll and Halsted was leveled the floor remained, exposing this short section of two-foot gauge track.

Bruce G. Moffat

Looking east on Randolph at Wabash in October 1992. Fiber optic conduits have been installed overhead. A former Tunnel Company wheel and axle set provides a silent reminder of the trains that once passed through this intersection.

Bruce G. Moffat Collection

One of the 535-554 series "homebuilt" Baldwins poses along side merchandise car 5162 and one of the illuminated safety signs in the late 1920"s.

In reality, the city bureaucracy had no interest in using the tunnels for any transportation purpose and probably wished they would just go away. In all likelyhood, the costs and logistics associated with providing vertical access from street level, building passenger stations and widening the tunnels to accommodate the related infrastructure would have been prohibitive.

Probably the most bizarre suggestion came from then-Cook County Sheriff Joseph I. Woods in 1968. He opined that the tunnels would be an ideal place to detain the many demonstrators who converged upon the city for the Democratic National Convention that year.

A Use Is Found

The first practical re-use of a tunnel segment actually occurred just four years after the last train ran. In late 1962 and early 1963, the city installed a 36" water main in the Michigan Avenue-Rush Street river crossing. Use of the tunnel saved an estimated $500,000 in construction costs. To access the tunnel, construction shafts were sunk at Illinois and Rush Streets and at the lower level of Michigan Avenue and South Water Street. The 16-foot long pipe sections were moved into place using a specially designed hand propelled car that ran on the two-foot gauge tunnel track.

Aside from this installation and a few minor steam heating pipelines mentioned earlier, the tunnels remained a resource in search of a purpose. That

Bruce G. Moffat

Looking south on Wabash from Randolph, one can discern the three spurs track connections that served Marshall Field's flagship department store. The temporary lighting in the 1992 view was for the benefit of work crews installing fiber optic cables.

purpose was found in 1972 when the Commonwealth Edison Company was looking for a way to provide the city's central area with additional electrical capacity.

In July 1972, representatives from Edison made a walking inspection of the system to gain a general knowledge of the tunnels' condition and their suitability for utility use. The tunnels were found to be in generally good condition, however several segments of particular interest were found to be flooded and had to be pumped out. Subsequent inspections located weak spots and other areas where deterioration had taken place and where seepage was a problem. On the whole, however, the tunnels were deemed to be in good enough condition that their rehabilitation and re-use would be cost-effective.

Negotiations followed, and in 1974 the city was presented with a proposition that it could accept. The agreement allowed the utility to rehabilitate six miles to house high-voltage conduits to serve the growing electrical demands of the Loop area. The use of the old freight tunnels saved Edison a substantial amount of money by avoiding the need to dig miles of trenches in streets already filled with pipes and conduits.

To provide access, shafts were built at various locations. Connecting tunnels were even built in a few places to provide continuity where the original tunnels had been destroyed by subway or building construction. These new tunnel sections, although of quite limited length, were generally built to the same dimensions as the originals but with steel liner plates to establish and sustain the required shape. The conduits consisted of two concrete encased ductways along each side of the tunnel.

The late 1980s saw Edison joined by a fiber-optic company and a cable television concern which found the tunnels ideal for housing their conduits. In succeeding years, additional mileage has been turned over to utility use.

Bruce G. Moffat

After the end of operations, city inspection crews walked the main portions of the system on an occasional basis. In 1980, the author, joined by transit writer Richard Kunz (at left) and two inspectors from the Department of Public Works, undertook one of several extended "expeditions" to photograph the remnants of this most unusual railroad.

A Walk "Forty Feet Below"

Fortunately, the author was presented with several opportunities over the years to visit the tunnels for the purpose of documenting their history and the role that they played in the city's commercial life. These opportunities were used to make physical inspections for the purpose of verifying prior research, determining the extent of additional tunnel mileage lost since the 1950s due to the construction of buildings and roadways, and to also document the extent of what had survived the salvage process. It was found that some mileage had been lost due to the building of the Grant Park underground garage on Michigan Avenue, as well as construction of the Sears Tower on Franklin Street and the Kennedy Expressway near Halsted Street.

Since research trips into the tunnels by the author were limited, the most was made of them. Walking trips of the various segments lasted for as long as five or six hours in the hopes that something interesting would be found. Stumbling into the "roundhouse" (repair shop) beneath Union Station was probably the single most notable example. More often, the miles of walking yielded only the occasional freight car, sign, or closed building entrance (curiously a late 1970s trip found the entrance into a Loop bank building to be wide open).

In most cases the tunnels had remained comparatively dry and in relatively good physical condition with few cracks evident. Spalling was noted in various locations and seepage at floor level was common. Contrary to some published reports at the time, inspections failed to confirm reports of garbage accumulation and its concomitant rodent infestation.

Bruce G. Moffat Collection

Bruce G. Moffat

The company's locomotive maintenance shop adjacent to Union Station was hardly a round house in the traditional railroad sense of the word. This sign was found on the wall of the Canal tunnel near Monroe where a spur track lead to the one-time mail handling facility.

On June 2, 1992, the author visited the Field Museum's spur for the filming of a documentary about the railroad "forty feet below." The museum's connection was untouched by the infamous "Loop Flood" of April 15th.

Bruce G. Moffat

A three-legged ladder, work benches, fixtures and assorted litter were found in the locomotive shop when first visited by the author in 1980. In the background is the narrow gauge inspection pit and overhead crane.

Although not as elaborately equipped as the locomotive shop, the nearby car facility still had intact but non-functional light fixtures and wiring.

Bruce G. Moffat

Bottom left: *Photographed in 1980, the severe deterioration evident on merchandise car 2417 resulted from an earlier flood of the then-inactive Union Station car shop in the early 1960's.*

Bottom right: *One of the Commonwealth Edison substations that supplied the 250 volt current.used for the trains was located at 10 E. Lake Street. In the adjacent tunnel there remains this power disconnect panel.*

Bruce G. Moffat

Bruce G. Moffat

Bruce G. Moffat

On Franklin between Washington and Madison Streets, there were two short trunk tunnel segments. This view of the north trunk tunnel looks south towards the connection with the south trunk tunnel (in the right background). The small opening directly ahead is actually a vestige of the original construction shaft #3 and for some reason was never enlarged to permit trains to pass through. Two of the post cards reproduced elsewhere in this volume were shot in the identical looking south trunk tunnel.

The Union Station shop was almost unchanged since abandonment, with overhead lights and hoists (not working, of course), lockers and work benches still in place along with a two-foot gauge inspection pit (used for locomotive maintenance). Evidence of at least one major flood sometime not long after the 1959 abandonment could be found on the stained walls. In between the two large bores that constituted the shop area sat a steel stake-sided merchandise car that had been missed during the salvage operation. The car exhibited evidence of that earlier flood in the form of major rusting on the lower four feet of the car but not above, indicating that the flood waters in this area did not completely fill the bores.

Bruce G. Moffat

For decades Baldwin-built locomotive 508 remained in the Field Museum's spur track until relocated to the Illinois Railway Museum in 1996. This view dates from 1980.

Preservation

As noted earlier, some rolling stock was overlooked by the scrappers and remained trapped underground. These included two flat cars in the Steele-Wedeles building in Dearborn Street, one locomotive and five loaded ash cars in the Field Museum's spur beneath Burnham Park and an assortment of cars and another locomotive west of the river. In 1983, Illinois Railway Museum volunteers removed the two Steele-Wedeles cars and set them aside at their museum near Union, Illinois, for future restoration.

Hope still lingered, however, that some day a means would be found for the railway museum's volunteers to remove Baldwin-built locomotive 508 and its five loaded ash cars from the Field Museum's spur. The train's existence was hardly unknown. In fact, it could be accessed directly from the Field's boiler room and over the years it was visited on a regular basis by both curious museum employees as well as the occasional rail enthusiast who had been fortunate enough to obtain permission to visit the boiler room.

Unfortunately, removal of the train was a complicated situation. Removal through the engine room was not practical because the tunnel's entry point into the basement was now obstructed by pipes and other equipment. The only other alternative would be to excavate from the surface – a costly proposition which the Illinois Railway Museum could ill-afford to shoulder alone given its limited financial resources. The cost of excavating the 12 feet to reach the tunnel coupled with a requirement to restore the tunnel (which also housed a water pipe) and the adjacent park land appeared to make the train's "rescue" impossible. Then,

The operator's position was still surprisingly intact although the inevitable corrosion was taking its toll.

Bruce G. Moffat

in 1996, opportunity knocked.

The city and the Chicago Park District were in the process of relocating a portion of Lake Shore Drive to improve traffic flow. As it turned out, the relocated roadway would pass directly above the spur's abandoned elevator shaft. Discussions between the Illinois Railway Museum and city to remove the train began late 1995 and continued into 1996. As part of the construction work, the cap protecting the shaft would probably have to be opened, making retrieval possible at a reasonable cost.

During the spring and early summer a host of logistical and legal issues were resolved including determining the railway museum's share of the construction contractor's expenses. The railway museum also would have to make arrangements for disposal of the cars' contents by a licensed hazardous material disposal company before they could be moved. Although one car held only ash, the other four cars were filled with asbestos, a hazardous commodity once used for insulating purposes).

At one point, hopes were nearly dashed when it appeared that it might not be necessary as part of the roadway relocation to remove the shaft's cap after all. This would have ended all hopes of obtaining the train. However, it was eventually determined that the shaft would have to be opened after all. Wasting little time, the railway museum's officers sprang into action, and through close coordination with city department of transportation officials, their engineering consultants and the Walsh Construction Company, arrangements were finalized to remove the train. Removal of the locomotive and cars would have to be handled with great care because of their deteriorated condition. This

Robert Rodenkirk

On August 7, 1996, Illinois Railway Museum and Walsh Construction Company personnel observe the removal of locomotive 508 from the Field Museum spur where it had sat for over forty years.

Robert Rodenkirk

The ash cars had to be unloaded before they could be raised out of the Field Museum's spur for transport to the Illinois Railway Museum. The fragile condition of this car is quite evident.

Robert Rodenkirk

Once out of the ground, locomotive 508 was quickly loaded into a truck for the trip to its new home.

was especially true of the 1906-vintage ash cars which had significant amounts of rotted wood and frames that were described by one observer as having a Swiss cheese appearance.

Finally, all was in readiness. On August 7, 1996, a steel cable was attached to locomotive 508 which was pulled 300 feet from where it had sat for over 40 years to the elevator shaft where it was placed on a steel plate sling and raised to the surface by a Walsh company crane. Alerted in advance, the press was on hand to record the engine's emergence into daylight for perhaps the first time since it was delivered to the Illinois Tunnel Company nine decades earlier.

The following day saw ash car 532 removed from the tunnel on its way to Union to join the locomotive. Retrieval of the remaining four cars which had been loaded with asbestos (653, 714, 766 and 856) had to wait a few days so that the hazardous waste contractor could complete his work. Using a winch, the still-coupled cars were towed to the elevator shaft and lifted onto a semi-trailer truck for their trip to the railway museum. Following a brief period on display, the locomotive and cars were placed in storage pending eventual restoration.

Preservation and Restoration

Bruce G. Moffat

During 2001, restoration of locomotive 508 by the Illinois Railway Museum's staff had begun in earnest. Seen here in the museum's buildings and grounds maintenance building, this venerable locomotive has already been sandblasted and given a protective coating of red primer.

Bruce G. Moffat

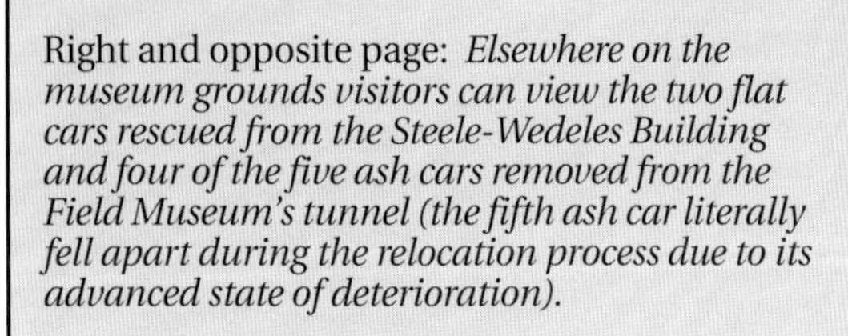

Right and opposite page: *Elsewhere on the museum grounds visitors can view the two flat cars rescued from the Steele-Wedeles Building and four of the five ash cars removed from the Field Museum's tunnel (the fifth ash car literally fell apart during the relocation process due to its advanced state of deterioration).*

Bruce G. Moffat

Bruce G. Moffat

CERA Goes "Forty Feet Below"

When the system was still operating the Central Electric Railfans' Association (CERA) made an overture to the Chicago Tunnel Company for permission to conduct an "inspection trip" (also known in rail enthusiast circles as a "fan trip"). Management was not interested and simply ignored the group's request. Any hopes to try at a later date seemingly ended with the operation of the last train in 1959.

When the author made his initial visits to the tunnels in the late 1970's and early 1980's permission was still difficult to obtain regardless of the stated purpose. Eventually however, conditions improved and visits for a narrow range of purposes began to be allowed by the city. Most of these were by utility concerns that were negotiating agreements to install conduits in the bores, although the occasional visit by news and documentary film crews were also accommodated.

In 2000, an inquiry by CERA resulted in a round of negotiations with the city's Department of Transportation culminating in an agreement allowing the organization to conduct a series of supervised one-hour walking tours for its members on September 1, 2000. Although there were no trains to photograph, members were shown the new bulkheads and water tight doors that had been installed following the 1992 Loop Flood, as well the fiber optic conduit installation and pumping system.

Phil O'Keefe

Looking east in the Randolph tunnel at Wells. The installation of the fiber optic conduits overhead meant that members had to wear helmets and stoop slightly to avoid contact.

Phil O'Keefe

Tunnel researcher Ed Anderson passes though one of two water tight "submarine" doors that were installed in the Randolph river crossing following the 1992 Loop Flood.

Bruce G. Moffat Collection

Prior to entering the tunnels, the tour itinerary is reviewed by trip director Bruce Moffat while the participants don boots and helmets.

11 The Loop Flood

Monday, April 13, 1992, began as just another typical workday for most Loop-area workers. However by lunch time many thousands of these same people would find themselves on an unexpected holiday - one that would last for days, if not weeks.

Less than a year earlier the Great Lakes Dredge & Dock Company, working under contract for the City of Chicago, had replaced a series of wood pilings that protected the Kinzie Street drawbridge from boat traffic on the Chicago River. Aside from having to place one of the replacement pilings in a slightly different location from the originals the work was completed without apparent incident. Then early on April 13, 1992, the engineers at several Loop-area buildings noticed rising water in their basements.

The first call to the city's 911 emergency center was received at 5:57 a.m. from the Merchandise Mart where two inches of water was found covering the basement floor. By the time firefighters arrived, the water had risen to two feet. By afternoon the water was 17 feet deep. At Marshall Field's and Carson Pirie Scott, the lowest two sub-basement levels were reported under water.

Within hours of the 911 call, the morning radio "drive time" programs were abuzz with speculation as to the source of the rising water that was rapidly swamping the lower levels of various, mostly older, large buildings including the Chicago Board of Trade, Pittsfield Building and City Hall. Water was

Bruce G. Moffat

The Kinzie Street bridge as it looked in April 1992. Note the pilings protecting the bridge house, which precipitated the most unusual floods ever.

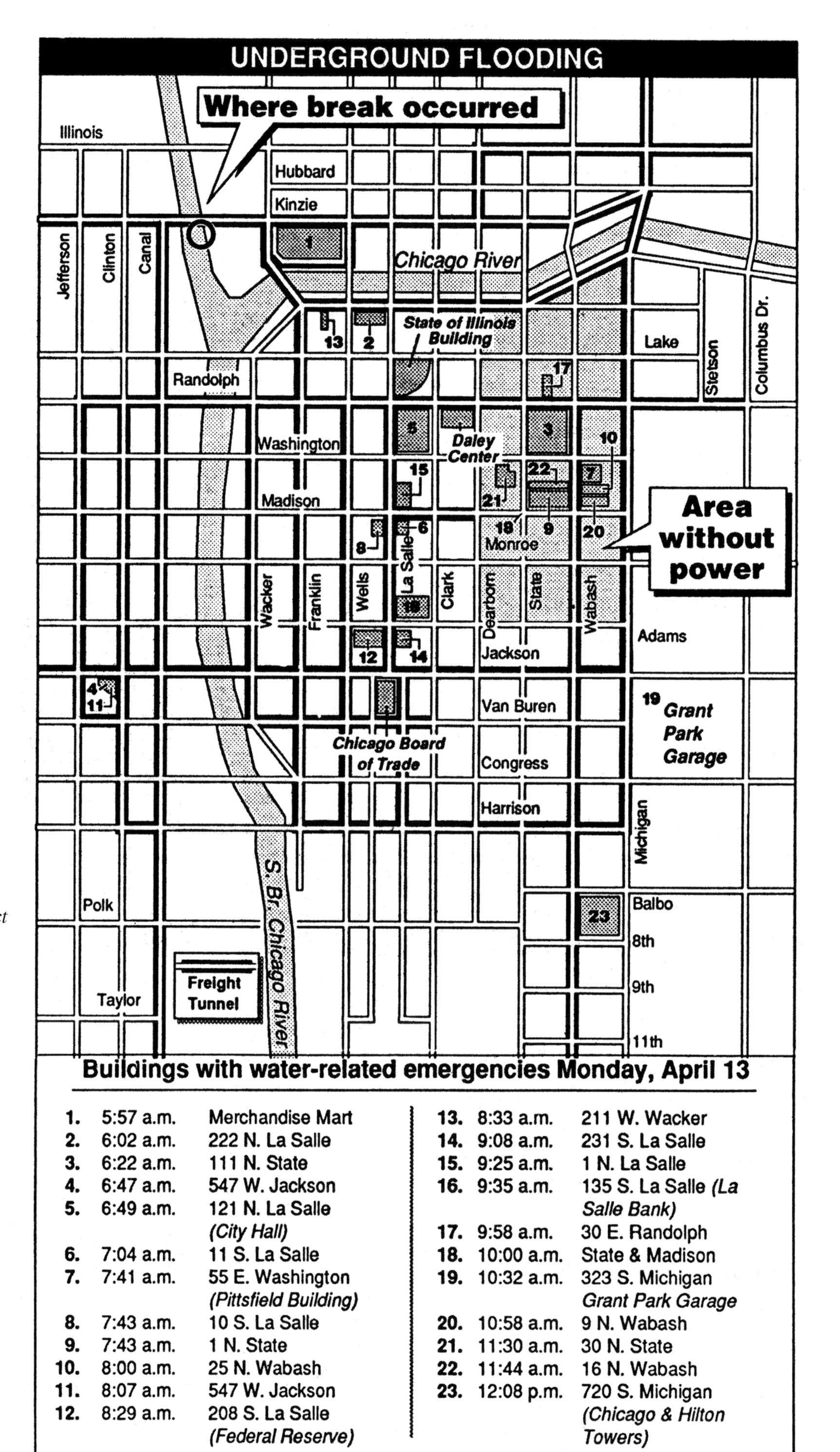

Buildings with water-related emergencies Monday, April 13

1.	5:57 a.m.	Merchandise Mart
2.	6:02 a.m.	222 N. La Salle
3.	6:22 a.m.	111 N. State
4.	6:47 a.m.	547 W. Jackson
5.	6:49 a.m.	121 N. La Salle *(City Hall)*
6.	7:04 a.m.	11 S. La Salle
7.	7:41 a.m.	55 E. Washington *(Pittsfield Building)*
8.	7:43 a.m.	10 S. La Salle
9.	7:43 a.m.	1 N. State
10.	8:00 a.m.	25 N. Wabash
11.	8:07 a.m.	547 W. Jackson
12.	8:29 a.m.	208 S. La Salle *(Federal Reserve)*
13.	8:33 a.m.	211 W. Wacker
14.	9:08 a.m.	231 S. La Salle
15.	9:25 a.m.	1 N. La Salle
16.	9:35 a.m.	135 S. La Salle *(La Salle Bank)*
17.	9:58 a.m.	30 E. Randolph
18.	10:00 a.m.	State & Madison
19.	10:32 a.m.	323 S. Michigan *Grant Park Garage*
20.	10:58 a.m.	9 N. Wabash
21.	11:30 a.m.	30 N. State
22.	11:44 a.m.	16 N. Wabash
23.	12:08 p.m.	720 S. Michigan *(Chicago & Hilton Towers)*

Map of the city's Loop district identifying flood-impacted buildings on the first day of this most unusual disaster.

When the flood hit, water flowed from the Wabash tunnel into the second subbasement of Marshall Field's through this doorway. When this photo was taken in December 1992, work was already in progress to seal this opening.

Bruce G. Moffat

also discovered leaking into the State and Dearborn subways and the Grant Park (south) underground garage. Eventually, the water came bubbling up into the lanes of the Kennedy Expressway in the area known as "Hubbard's Cave." At first it was thought that there had been a major water main break, but the rate at which the water was rising ruled out that possibility.

At first no one, except possibly the by-now-frantic maintenance staffs at the affected Loop buildings, knew the source: the largely idle and half-forgotten freight tunnels. Other, mostly newer, buildings which had never been connected to the system were, with a few exceptions, unaffected. City officials began ordering the evacuation of most downtown office buildings. The rising water was beginning to endanger underground electrical distribution facilities, raising the possibility of power failures that would leave high-rise office workers stranded. As a safety precaution, Commonwealth Edison shut off power to a large portion of the downtown area, forcing stores and office buildings to close. The Chicago Board of Trade and the Chicago Mercantile Exchange were forced to suspend trading. Banks closed their doors and hurriedly implemented disaster management plans.

The "flood" also forced the Chicago Transit Authority to suspend service through the State Street subway and reroute those trains via the Loop Elevated at about 10:25 a.m. Lacking an alternative route through the Loop, trains using the Dearborn Street subway were turned back at Division on the north (O'Hare branch) and at Racine on the south (Congress and Douglas branches). A shuttle bus service was hastily arranged to carry passengers between those stations and the Loop. (The State Street subway would not reopen until May 1, with the Dearborn Street subway following suit on May 7.)

By late afternoon the Loop area was a virtual ghost town. Large portions of the Loop were completely dark. Many stores and offices, especially those along State Street and Wabash Avenue, would remain closed for days, if not weeks, due to the interruption of power and flooded engine rooms. Power to a 26-block area of the Loop was shut off as a safety precaution as the water rose to cover underground electrical transformers and high voltage cables.

Loop banks and the financial houses managed to resume at least some functions within a few days. Marshall Field's department store reopened after one week thanks to an improvised electrical supply hookup, but without working air conditioning. Reportedly, the water was still 35 feet deep in the building's subbasement levels. Even after the water was pumped out, repairing the damage would go on for months. Down the street, Carson Pirie Scott reopened a week later.

Electronic and print media reporters and photographers from around the world descended on Chicago to record this most unusual flood – a process complicated by the fact that the water and its resulting damage were hidden from easy view. Network television news programs began with the latest flood update from Chicago. The concept of an "invisible" flood baffled more than a few viewers.

Chicago Sun-Times

On April 14, 1992, stone was dumped into the river in an attempt to stop the flooding.

Something Fishy at City Hall

The saturation-level flood coverage included its share of offbeat stories. Two days into the flood the Chicago *Tribune* reported that a 1970 prediction made by the late Mayor Richard J. Daley, father of the current mayor, had, in a rather odd way, come true:

> It was June of 1970 when Mayor Richard J. Daley announced his dreams for the Chicago River: "I hope to see the day there will be fishing in the river...perhaps swimming."
>
> Many people laughed and one newspaper reported there had not been a live fish seen in the river for four years.
>
> But Daley persisted, and in 1973 he told visiting congressmen that he envisioned Loop workers spending lunch hours catching and grilling fish on the river's banks.
>
> Tuesday, workers needed only to visit the flooded sub-basement of City Hall to find fresh fish flopping about, courtesy of the flood waters that coursed through the Loop.
>
> Sam Dennison, a biologist with the Metropolitan Water Reclamation District, was not surprised at the stock because millions of dollars have been invested in cleaning up the river over the last two decades...
>
> A few blocks away in the Randolph Street Metra Station engineering supervisor Jim Simpson held a jar containing a lake perch that had been fished out of the basement...
>
> "I could not even guess how many fish were sucked into the hole," Dennison added, "but if anyone finds one with our yellow anchor tag on it, we would like to hear from them."

Fighting the Flood

Within hours, the location of the water's penetration was determined to be adjacent to the bridge pilings that protected the south side of the Kinzie Street bridge house near the east bank of the river. Passers-by looking over the bridge railing could readily observe the small whirlpool, which signaled the location where the river was draining into the tunnels. Meanwhile, many buildings hurriedly rented pumps in a largely futile attempt to combat the still rising water and limit

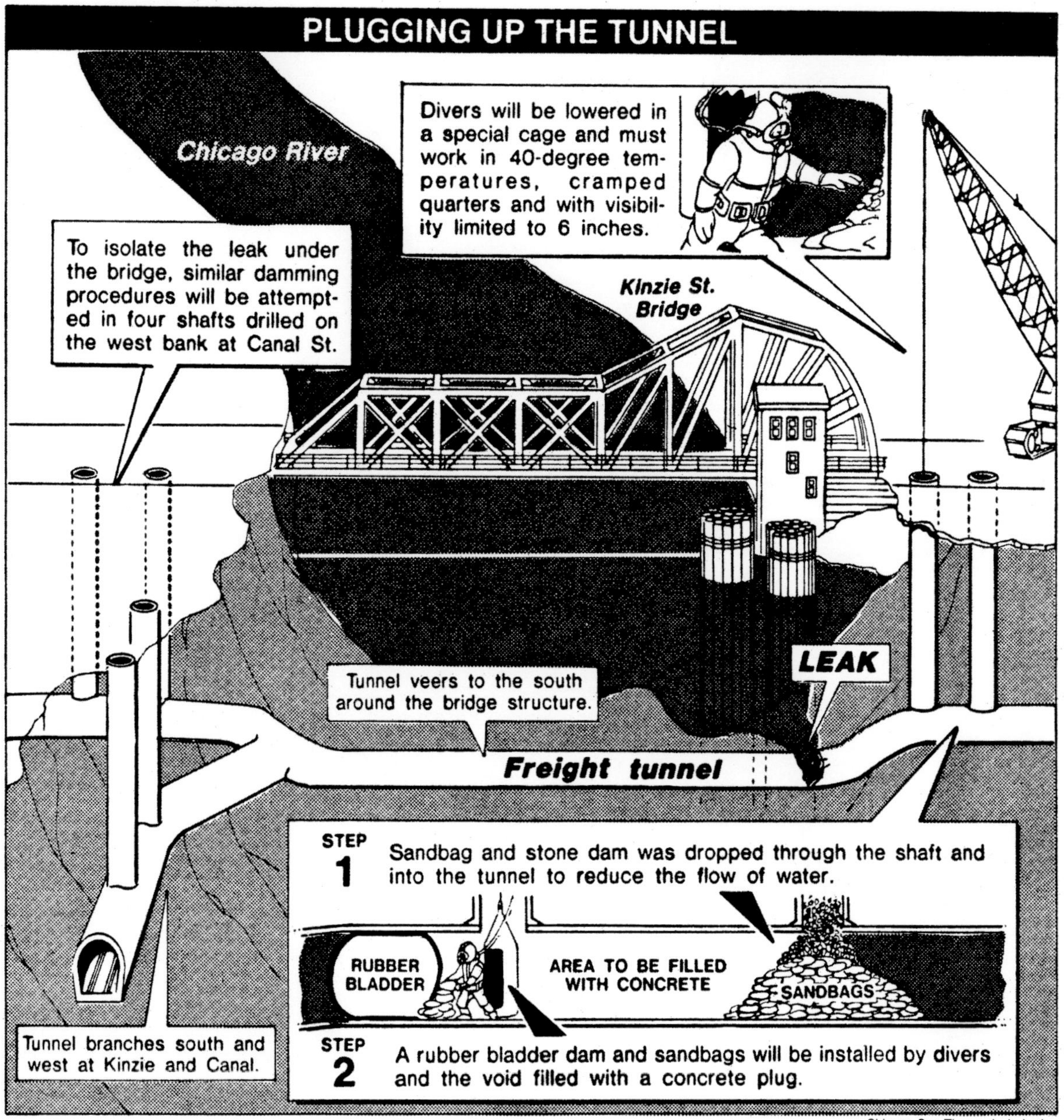

Newspaper graphic explaining the technique that was ultimately used to stop the water flow and seal the damaged tunnel segment.

the consequent damage. Many of these buildings had three subterranean levels and open access ways. Others that had installed bulkheads years earlier found that these plugs were not watertight.

Workers at the County Building rushed to rescue documents, some more than 100 years old, from the rapidly intruding water. At Orchestra Hall, workers hurriedly relocated instruments and an entire music library out of harm's way. At many buildings, maintenance staff could only look on helplessly as the rising water seriously damaged heating, ventilation and electrical equipment. At the Chicago Board of Trade, a CBOT executive answered a reporter's query with a statement that sounded more typical of a suburban homeowner dealing with a sump pump failure: "I can't talk to you now, I'm trying to get the water out of my basement!"

During the day, loads of gravel and sandbags were rushed to the site and poured into the whirlpool in an effort to plug "the leak," which was now estimated to be the size of a small automobile. A load of mattresses was even brought to the site but it was decided not to use them. Despite these efforts, the water continued rising in building basements at the rate of about two feet an hour. By noon, 23 buildings had reported flooding. Ultimately, a special quick drying concrete was pumped into the hole, slowing the flow of water to a relative trickle. The tunnels were now almost completely flooded.

Mayor Richard M. Daley and Illinois Governor Jim Edgar held news conferences to assure the public that all possible steps were being taken to contain the damage and to determine the chain of events leading up to the failure of the tunnel wall. In the next morning's edition of the Chicago *Sun-Times*, there

Jack Higgins. Used by permission.

Drawing Their Own Conclusions

The flood's novelty was not lost on local, and even a few national, political and popular cartoonists who wasted no time in giving Chicago's unofficial motto, "The City That Works," a thorough drubbing. One nationally-syndicated cartoonist drew a parallel with the infamous Chicago fire of 1871 by suggesting that the flood may have been instigated by a descendant of Mrs. O'Leary's cow. Local lore incorrectly attributed the blaze, which destroyed a large portion of the city, to a cow knocking over a lantern in a barn behind O'Leary's home.

A cartoonist for a national railroad magazine took a completely different approach and showed a group of workers inspecting the tunnel after the flood waters had receded and discovering a New York, Susquehanna & Western Railroad steam locomotive that had apparently been sucked in by the whirlpool. The actual locomotive, which had been built in China, had earlier been aboard a ship enroute to the east coast tourist line when it sank in the Indian Ocean.

In a tunnel under the Chicago River, a descendent of Mrs. O'Leary's cow follows her calling.

The Far Side ©1992 Farworks, Inc. Used by permission. All rights reserved.

RAILFUN

by CURT KATZ

YES, MISTER MAYOR, WE'VE FINALLY GOT MOST OF THE FLOOD WATER PUMPED OUT OF THOSE OLD CHICAGO TUNNEL COMPANY FREIGHT SUBWAYS...

CHICAGO "THE CITY THAT WORKS"

MEMOS

Curt Katz/Carstens Publications Used by permission.

Jack Higgins. Used by permission.

Chicago Sun-Times

Bruce G. Moffat

appeared a status report about the flood that included some of the first indications that the city officials were aware of the Kinzie tunnel's weakened condition:

> Much of Chicago's central business district will be without power today as crews try to repair the damage from a freak underground flood that sent Chicago River water into some of the city's most important basements.
>
> Mayor Daley said late Monday that the leak in a turn-of-the-century underground tunnel system where it crossed beneath the river at the Kinzie Street bridge was "partially reduced."
>
> An admittedly angry mayor conceded at a 10 p.m. briefing that some city employees had known about the leak before it shut down the Loop. "There were people who had information about a minor leak in the system," he said. "There was information within the system, not in my office. There was information filed, reports filed."
>
> Singling out the Public Works department, Daley said, "Individuals did drop the ball. The city didn't. Individuals should be held accountable."

Later that Tuesday, the mayor demanded and received the resignation of the acting public works commissioner for not expediting repairs to the tunnel when the first reports of damage reached his department. Other resignations and firings quickly followed as the mayor sought to limit the political and public relations damage, and perhaps potential liability claims. Corporation Counsel Kelly Welsh was directed to investigate and determine the circumstances that led to the city's greatest disaster since the famous Chicago Fire of 1871. Meanwhile, the press had a field day covering serious and lighter aspects of the most unusual disaster to strike a city anywhere.

To repair the rupture, the city turned to the Kenny Construction Company, an area firm with extensive experience in large construction projects. Work on plugging the tunnel was personally supervised by the family-run company's vice president of operations, John Kenny, whose professionalism and demeanor quickly made him a media favorite. The mayor's hiring of Kenny also brought an overwhelming positive response from business and building owners. Supervision of the "dewatering" process was left to the U.S.

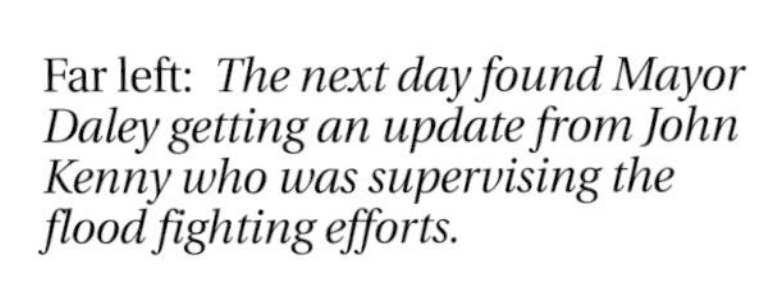

Far left: *The next day found Mayor Daley getting an update from John Kenny who was supervising the flood fighting efforts.*

Left: *When the tunnels were built in the early 1900's, manually operated "flood doors" were installed at river crossings. The rusted remains of this long inoperable door on the approach to the Wells Street river crossing was photographed not long after the 1992 Loop flood.*

Right: *A few days after the flood tunnels flooded, the city released this photograph to local media outlets. Taken in March 1992, it clearly shows the extent of the damage.*

City of Chicago

Army Corps of Engineers which had jurisdiction over waterway and flood control projects.

A flood command center was set up in a former factory building near the Kinzie bridge site to coordinate the repair and dewatering process. As a part of this work representatives of the city, Kenny, the Army Corps of Engineers and other agencies conducted daily press briefings. There was no doubt that the city had acted proactively by marshaling every available resource to cope with the situation.

A determination was quickly made that abandonment of the damaged tunnel segment was a more practical solution than reconstruction. Accordingly, shafts were sunk on either side of the river to allow divers to enter the flooded tunnels, assess their condition, and monitor the flow of water. On Thursday, April 16, a diver was lowered into the water-filled tunnel and reported that the water was now flowing at a fairly slow rate of 0.3 feet per second. Visibility was limited to 6 inches, however. Conditions were now deemed suitable to construct the temporary and permanent bulkheads. Only after this dangerous work was completed could the tedious dewatering process get underway.

By Friday, April 17, one estimate placed flood-related losses (both physical damage and lost business due to forced closures) at a staggering $1 billion. Meanwhile, crews labored to completely stop the water flow. Only a day earlier, the U.S. Army Corps of Engineers had estimated that it could take 15 days to plug the hole and drain the tunnels – an estimate that proved to be remarkably accurate.

The city's investigative team soon determined that the installation of the new pilings at the Kinzie Street bridge had forced some existing submerged timbers from an earlier piling to pierce the tunnel wall. This damage started a slow leak allowing silt from the river bottom to slowly enter the tunnel months before the collapse. On April 22, ten days into his investigation, Welsh appeared at a press conference to present the following chronology of events leading up to the flood:

December 19, 1990: The city advertised for bids to replace wooden pilings protecting five Chicago

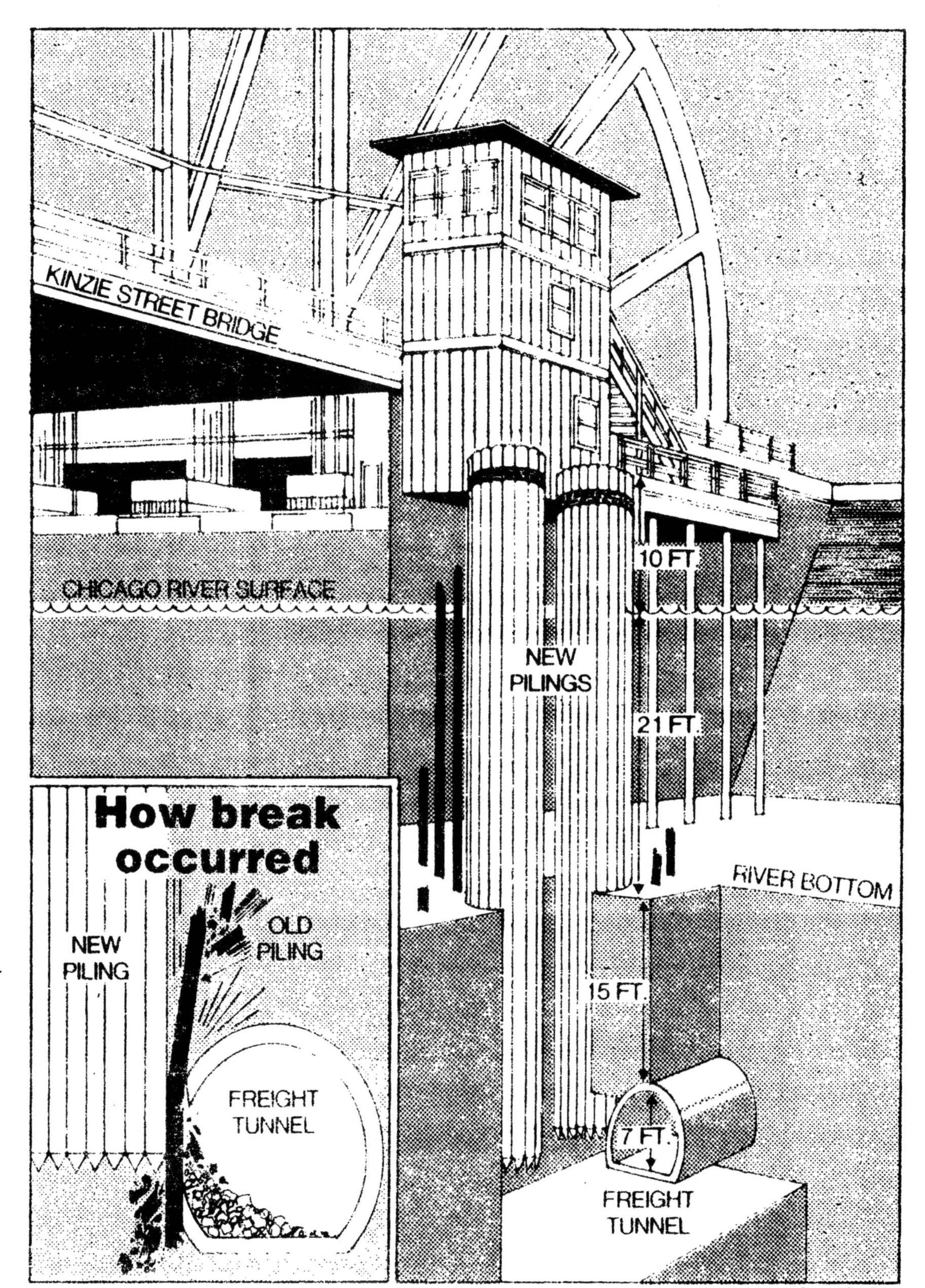

This graphic illustrates how the replacement piles punctured the wall.

River bridges, including the one at Kinzie. On January 15, 1991, Great Lakes Dredge & Dock Co. submitted a low bid of $335,650.

August 26, 1991: Great Lakes began work at the Kinzie bridge. The city had not warned Great Lakes of a tunnel under the bridge, but the contract required the firm "to inform itself" of underground structures nearby before it sank the pilings.

September 11, 1991: Great Lakes explained to the city's resident and project engineers that two piling clumps south of the bridge had to be slightly relocated "for logistical reasons." (According to published news reports, the project engineer later said that he had verbally approved the location changes and "expected a written change order [request] later," as procedures normally require. Nothing written was ever submit-

Bruce G. Moffat

A short distance from the ruptured tunnel was this pump which had been used until the late 1950's to keep the Kinzie river crossing and adjacent areas passable.

ted and he never checked the new sites with higher-ups. The resident engineer never mentioned the relocation in progress reports.)

October 11, 1991: Two city engineers met with Great Lakes officials at the Cermak Road bridge and "inspected the project" by checking work only at that site. One of the engineers was reported to have said later that the other work sites, including the one at Kinzie Street, weren't inspected because "there was nowhere to park" near them.

January 14, 1992: John Kohnke of Digital Direct, a cable TV consulting company, accompanied by a photographer, shot videotape of the breach while inspecting the tunnels for a possible cable installation. Kohnke said he tried once to call his city contact, James McTigue.

According to published news reports, Welsh's chronology went on to state that Kohnke made contact with McTigue on February 27. It was also reported that a March 13 notation in McTigue's diary indicated that the wall had collapsed and that he could not make a close examination due to the heavy silt, but that he had reported the problem to his supervisor. On March 18, he returned to the location to photograph the breach. The tunnel wall had been "pushed in" in an area six feet wide by two or three feet high. McTigue took the exposed film to a drug store for processing and picked-up the prints on March 25. He then met with the city's soil engineer and it was decided to submit the photos as soon as possible for review by the city's Department of Transportation.

During the first several days in April, memos were exchanged and meetings held to determine how to fix the tunnel and at what cost. A memo requesting permission to go ahead with the work was sent to acting Transportation Commissioner John LaPlante who responded on April 5 with a memo giving the go-ahead to contract for the needed repairs. The process to hire a contractor to make the repairs was then initiated.

And then the wall gave way.

Chicago Historical Society (ICHi-23686)

After the flood waters had receded, large fans and heavy-duty dehumidification equipment was brought in to dry out the lower levels of DePaul University's building on Wabash Avenue.

The Dewatering Process

On April 19, the ruptured tunnel was declared sealed-off and two days later the Army Corps of Engineers began pumping operations. Concerned about the potential effect that a rapid lowering of the water level might have on the tunnel's structural integrity, the rate was kept to a conservative 1 to 3 inches per hour. By April 26, the Corps had removed an estimated 117 million gallons. By the time the process was completed on May 21, an estimated 134 million gallons had been removed.

As the water level continued to drop and electrical service was restored, the 16 most-ravaged buildings gradually came back to life. On May 6, the 222 N. LaSalle Building (formerly known as the Builders

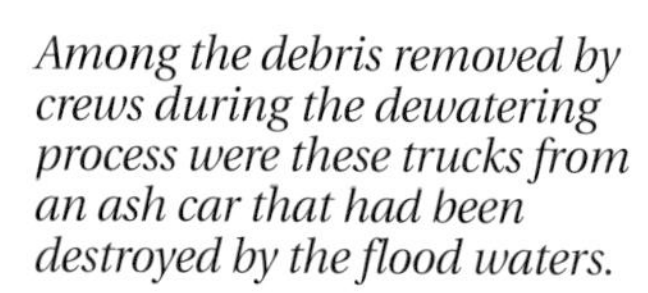

Among the debris removed by crews during the dewatering process were these trucks from an ash car that had been destroyed by the flood waters.

Bruce G. Moffat

Building) reopened with a catered party for its long-inconvenienced tenants. At the height of the flood, the nearly 70 year-old building had 34 feet of water standing in its lower levels.

Fortunately, none of the affected buildings experienced structural damage although electrical, heating, air conditioning and other building support systems were often damaged beyond economical repair and had to be replaced.

Once the dewatering process had been completed, Kenny Construction installed permanent bulkheads on either side of the breach. The city hired Harza Engineering to inspect the tunnels for damage, design bulkheads for the other river crossings, and develop recommendations for maintaining and managing the tunnel system.

On December 10, an ebullient Mayor Daley wrote "finis" to the Loop Flood, saying, "We're closing the books on one of the most dramatic chapters in Chicago's history." The declaration came as he announced the completion of six-foot thick concrete bulkheads at 13 locations where the tunnels crossed a waterway. Designed to keep the river from ever again invading the tunnels, bulkheads at seven of the crossings included watertight steel access doors and conduits to carry utility lines. The remaining six crossings were given solid concrete plugs.

Accompanied by John Kenny, the mayor and a small party of construction and engineering executives were lowered down a construction shaft near the Merchandise Mart to inspect the bulkheads protect-

Chicago Sun-Times

Cook County employee Mary Griffin examines flood-damaged documents in the County Building's basement.

Chicago Sun-Times

In mid-1992, Kenny Construction's John Kenny (left) and Mayor Richard M. Daley inspected one of the new bulkheads that were installed in the Kinzie Street tunnel.

Bruce G. Moffat

In June 1992, with the dewatering process largely completed, workmen had turned their attention to removing the large accumulations of silt and other debris. The location is the Kinzie tunnel just west of the former river crossing.

ing the Wells Street river crossing.

The decision to maintain the system's continuity in this manner arose from the realization that the tunnels could serve a practical purpose for the city. Daley touched on this point when he told reporters: "The flood reminded us all of what a unique civic asset the tunnel system is. We're going to market it hard, as a great communications pathway beneath the city." The system's profit potential had already been demonstrated. Previous installations in the tunnels by Commonwealth Edison, Illinois Bell Telephone and a fiber optic company were already netting the city a reported $10 million annually. Plans were also announced to install improved drainage and lighting systems to make the bores more "user friendly" for prospective utility installations.

Following the installation of the bulkheads, the tunnels gradually receded from the public's consciousness as Loop commerce returned to normal. Lawsuits resulting from the flood would require another eight years to resolve.

Bruce G. Moffat

A June 1992 visit found 3668 and several other merchandise cars sitting in about 18 inches of water on Clinton just north of Lake.

Bruce G. Moffat

The Flood Becomes A Museum Exhibit

Barely had the Loop Flood begun to recede from the public's memory then an exhibit looking back on the event opened at Chicago's Museum of Science & Industry in June 1992. Titled "Chicago Springs a Leak," the temporary exhibit included historical photographs, artifacts recovered during the dewatering process, diving equipment of the type used at the Kinzie Street site, and even some flood-theme T-shirts sold by sidewalk vendors in the days immediately following the April 13th wall collapse.

Bruce G. Moffat

Display of the type of equipment worn by divers while assessing the flow of water into the tunnels at the Kinzie Street site.

Bruce G. Moffat

Entrance to the flood exhibit was through an archway loosely patterned after the profile of the real bores.

Bruce G. Moffat

The skeleton of one of the ash cars that had rested not far from the Kinzie Street river crossing was recovered and put on display. This car was later donated to the Fox River Trolley Museum. Note the historical photographs displayed on the wall.

SECTION V Rolling Stock

12 Locomotives and Cars

Beginning in 1915, and continuing through 1959, both the Chicago Tunnel Company and the Chicago Warehouse & Terminal (later the Chicago Tunnel Terminal) Company filed annual reports with the Interstate Commerce Commission in Washington, D.C., as well as with the Illinois Commerce Commission in Springfield.

In addition to the usual profit and loss statements, these reports also included a tabulation of equipment owned. Although very superficial in nature, these reports provided the only details made generally available on a continuing basis about the size and nature of the fleet.

Larry Best Collection

Left: *Mail car #40 and a crew of loading dock workers pose for the company cameraman at the Illinois Central Railroad's Park Row Terminal (later known as Central Station) in 1906.*

Chicago Historical Society (ICHi-31227)

Above: *For bulky cargo, flatcars equipped with pole-type stakes and chains were used from 1906 until at least the 1920's. Photographed on the first day of revenue operations on August 15, 1906, car 3047 has obviously seen extensive use in construction service.*

EQUIPMENT OPERATED
CHICAGO TUNNEL COMPANY – 1915-1959

Source: Interstate Commerce Commission

Year	Freight	Work	Miscellaneous*	Locomotives	Passenger
1915	2,985	2	5	132	
1916	2,982	3	5	126	
1917	2,982	3	5	126	
1918	2,976	3	5	126	
1919	2,976	3	5	126	
1920	2,967	3	5	126	
1921	2,967	3	5	126	
1922	2,967	3	5	126	
1923	2,917	3	5	136	
1924	2,770	52		146	
1925	2,917	60		143	
1926	3,036	60		143	
1927	3,244	60		143	
1928	3,216	62		149	
1929	3,219	61		149	
1930	3,237	62		144	
1931	3,215	62		144	
1932	3,243	58		138	5
1933	2,528	46		129	6
1934	2,451	48		128	6
1935	2,405	48		128	6
1936	2,408	45		127	6
1937	1,985	45		119	6
1938	2,410	45		119	6
1939	1,910	45		119	6
1940	1,922	45		118	6
1941	1,665	46		106	6
1942	1,634	46		106	6
1943	1,583	44		86	6
1944	1,534	44		86	6
1945	*Figures not available*				
1946	1,617	18		90	5
1947	1,614	18		90	5
1948	1,592	18		90	5
1949	1,546	18		90	5
1950	1,527	18		90	5
1951	1,527	18		90	5
1952	1,474	18		90	5
1953	1,406	18		90	5
1954	1,401	18		83	5
1955	1,393	18		83	5
1956	1,387	18		83	5
1957	1,387	18		83	5
1958	1,361	18		83	5
1959	1,357	18		83	5

* Inspection cars were listed as passenger cars beginning in 1932. These cars not reported as a separate item 1924-1931, although several were on the property.

The first Jeffrey locomotive purchased by the Illinois Telephone Construction Company on behalf of the Illinois Tunnel Company carried fleet number 119 (serial #804). Initially equipped with this experimental "diamond" trolley, it was soon re-equipped with a conventional trolley pole.

Jeffrey Division, Dresser Industries.

The Locomotive Fleet

During the first several years of operation, most of the locomotives operated by the Illinois Tunnel Company were purchased through the subsidiary Illinois Telephone Construction Company, but were lettered for Illinois Tunnel. Aside from the Morgan cog-rail units described elsewhere in this book, the products of five builders comprised the electric roster: Goodman, Baldwin-Westinghouse, General Electric, Jeffrey and Whitcomb. Locomotives were originally painted olive green with black used later on. By the 1950's painting had been discontinued, apparently to save money. Rounding out the roster were several Porter steam locomotives and a handful of Baldwin gasoline locomotives.

The following table summarizes the locomotive fleet by builder and is organized according to the fleet numbering system in effect from 1909:

Fleet #	Builder	Date	Total Units
(4)	Morgan	1903-05(?)	4
(2)	Porter	?	2
200-225	General Electric	1905	26
301-305	Goodman	1908	5
400-467	Jeffrey	1904-1907	68
500-524	Baldwin-Westinghouse	1907-1908	25
525-534	Baldwin-Westinghouse	1913	10
554-554	company shops*	1923-1924	20
(1)	Whitcomb	1922	1

*Assembled from kits supplied by Baldwin-Westinghouse.

Morgan

The East Chicago, Indiana-based Morgan Electric Machine Company built at least four center third rail-powered cog locomotives that were primarily used for spoil removal during the early days. Little specific technical information is available on these units, which included three different models. Of the several photographs made of these locomotives, only two fleet numbers can be determined – 109 and 110; these were single motor units.

Fleet #	Serial #	Date	Motors	HP	Wt.
(2)	?	1903	1	80	6,000
(1)	?	1905(?)	2	?	10,000
(1)	?	1905(?)	1	80	6,000

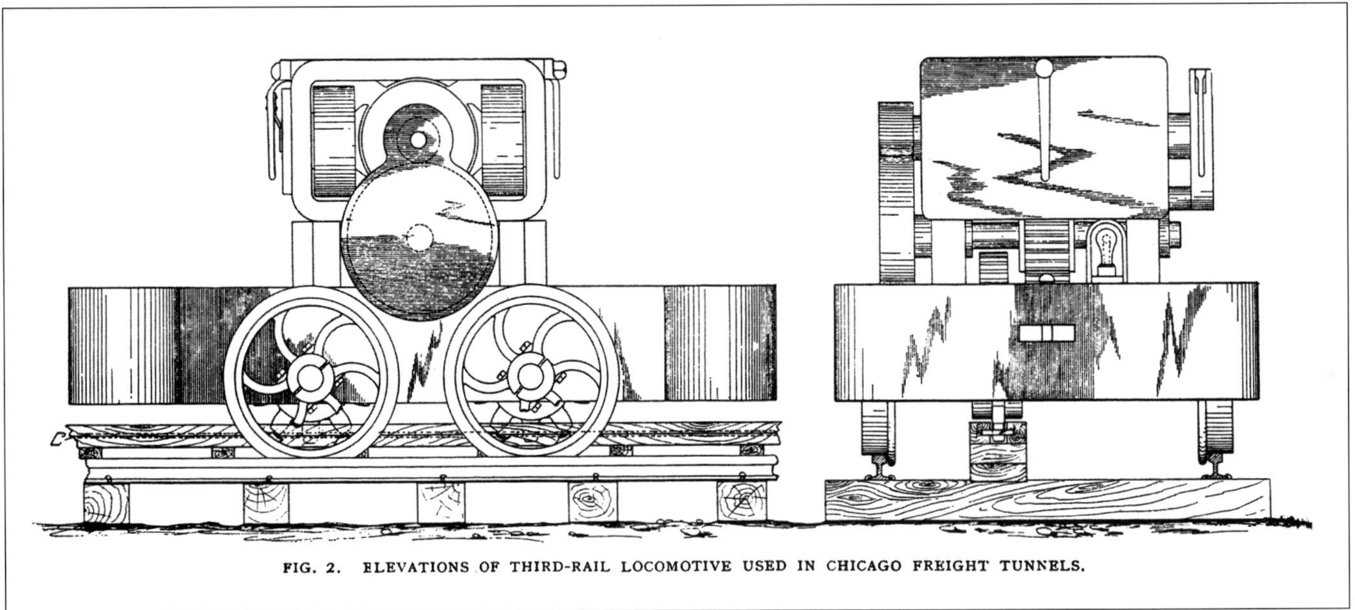

FIG. 2. ELEVATIONS OF THIRD-RAIL LOCOMOTIVE USED IN CHICAGO FREIGHT TUNNELS.

Western Electrician

Plan of one of the single motor Morgans. For some unknown reason, the third rail is incorrectly depicted as being off-center.

MORGAN ELECTRIC THIRD RAIL HAULAGE SYSTEM

MANUFACTURED BY

MORGAN ELECTRIC MACHINE COMPANY

EAST CHICAGO, INDIANA

80 H. P. Morgan Third Rail Locomotive Used in Freight Haulage in Tunnels under the Streets of Chicago.

No sand, trolley pole or trolley wire used. Positive working with light rails on all grades. This system is now extensively used by many of the largest producers of coal throughout the country. Let us tell you about it. Send for Catalogue "A A."

GOODMAN MANUFACTURING CO.

SELLING AGENTS

HALSTED ST. AND 48TH PLACE (NEAR STOCK YARDS

CHICAGO, U. S. A.

Goodman Equipment Corporation

One of the Tunnel Company's third rail locomotives was featured in this advertisement that appeared in the March 1904 issue of Electrical Mining.

Porter

Porter was well known for its line of industrial and narrow gauge steam locomotives. To assist in landfill operations at Grant Park, the Illinois Telephone Construction Company obtained two, apparently used, 24" gauge 0-4-0 saddle tank locomotives. Like all other rolling stock, these were lettered for the parent Illinois Tunnel Company. Photographs show that one of these was numbered 19. The locomotives were removed from the property when landfill operations were completed in 1907. No other information concerning their service history or mechanical specifications has been found.

Goodman

The Chicago-based Goodman Manufacturing Company built just five trolley-type locomotives. Delivered in 1908, and bearing road numbers 301-305, they came equipped with a single 40-horsepower motor, inside frames and a worm-drive system. They were not operating favorites and were out of service by 1910. At least three were resold to mining companies: 301 to the Applegate & Lewis Coal Company on April 23, 1910; and 304 and 305 to the Saylor Coal Company on December 1, 1910.

Fleet #	Serial #	Date	Motors	HP	Wt.
301-305	?	1908	?	40	?

Baldwin-Westinghouse

Based on court records and Commerce Commission filings, Baldwin-Westinghouse supplied at least 40 two-motor electric locomotives in "ready to run" condition. Baldwin gave their wheel arrangement as "mule" – a somewhat fitting description. At least a few units were originally numbered in the 100 and 200 series, but all were assigned 500 series numbers by 1909, leaving the 200 series to the GE's.

The company's roster also included 20 "home-assembled" units, which were supplied by Baldwin in kit form. Final assembly was performed in the company's shops during 1923-24. Since these units were not supplied "ready to run," the company elected to attach their own rectangular builders' plate to each unit identifying them as being built by the Chicago Tunnel Company. As late as 1958, all 20 were on the roster but were probably inoperable.

Fleet #	Serial #	Date	Motors	HP	Wt.	Comments
173	29749	12/1906	?	?	?	Disposition/renumbering unknown.
500	31873	10/1907	2-WH 155	30	?	Originally 205.
501:I	32308	11/1907	2-WH 155	30	?	Renumbered 508 (II)
501:II	31923	10/1907	2-WH 155	20	?	Originally 206.
502-504	32319-21	12/1907	2-WH 155	30	?	
505-507	32401-03	12/1907	2-WH 155	30	?	
508:I	32404	12/1907	2-WH 155	30	?	Renumbered to 509.
508:II	32308	11/1907	2-WH 155	30	?	Originally 501 (I)
509	32404	12/1907	2-WH 155	30	?	Originally 508 (I)
510-516	32552-58	01/1908	2-WH 155	30	?	
517-518	32569-70	01/1908	2-WH 155	30	?	
519-524	32629-34	01/1908	2-WH 155	30	?	
525	40938	11/1913	2-WH 903	50	14,000	
526	40939	11/1913	2-WH 903	50	14,000	
527-534	40981-88	12/1913	2-WH 903	50	14,000	
535-554	?	1923-24	2-WH 903 ?	?	?	Supplied as kits. 553 still in tunnel.
(5)	?	1925	2-WH 155	50	14,000	Status uncertain.

In addition to the electrics, Baldwin supplied four gasoline-engined locomotives. Although purchased in the name of the Chicago Tunnel Company, these units were assigned to the sister Chicago Warehouse & Terminal Company and used at the Burnham Park landfill site. These locomotives were apparently retired and off the property prior to 1925.

Fleet #	Serial #	Date	Motors	HP	Wt.
(2)	40322-3	08/1913	?	50	?
(1)	43330	05/1916	?	50	?
(1)	?	?	?	50	?

Plan of one of the first Baldwins to be shipped to the Illinois Tunnel Company.

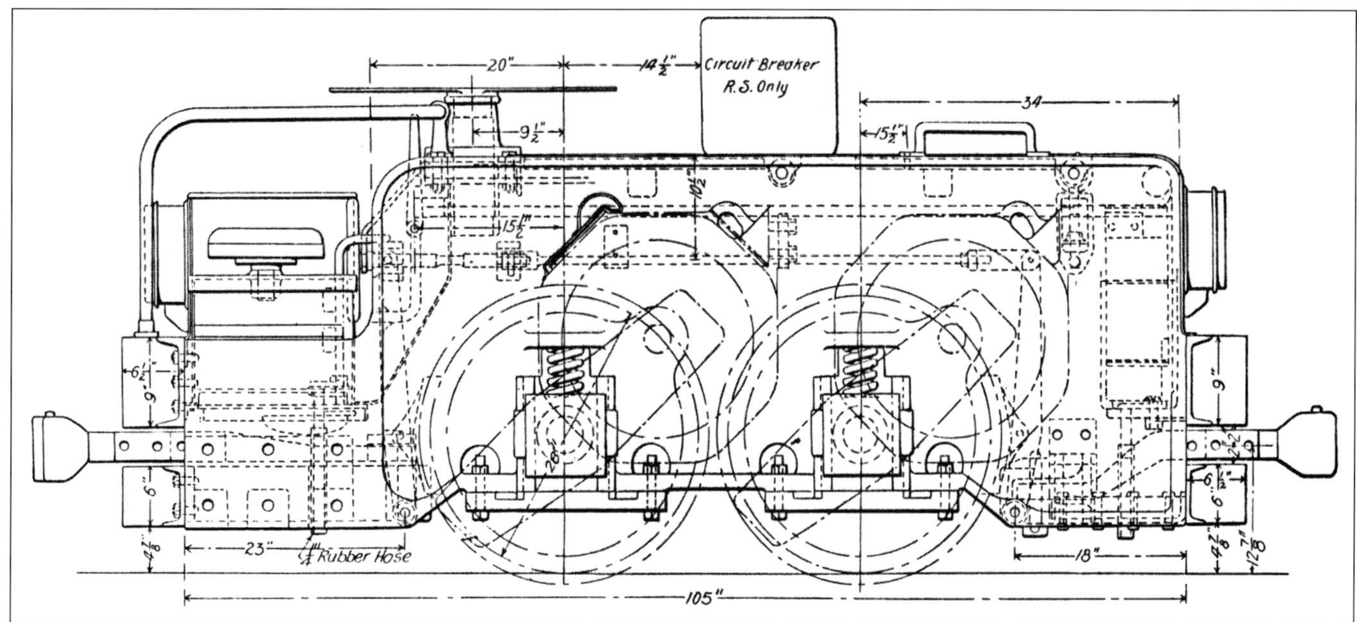

Bruce G. Moffat Collection

Locomotives 535-554 were supplied in kit form. So in lieu wearing the standard Baldwin-Westinghouse builders' plates, the Tunnel Company's shop forces installed their own markers. This brass plate adorned locomotive 553.

Bruce G. Moffat

Builder's photo of one of the first Baldwins to be shipped to the Illinois Tunnel Company.

Bruce G. Moffat Collection

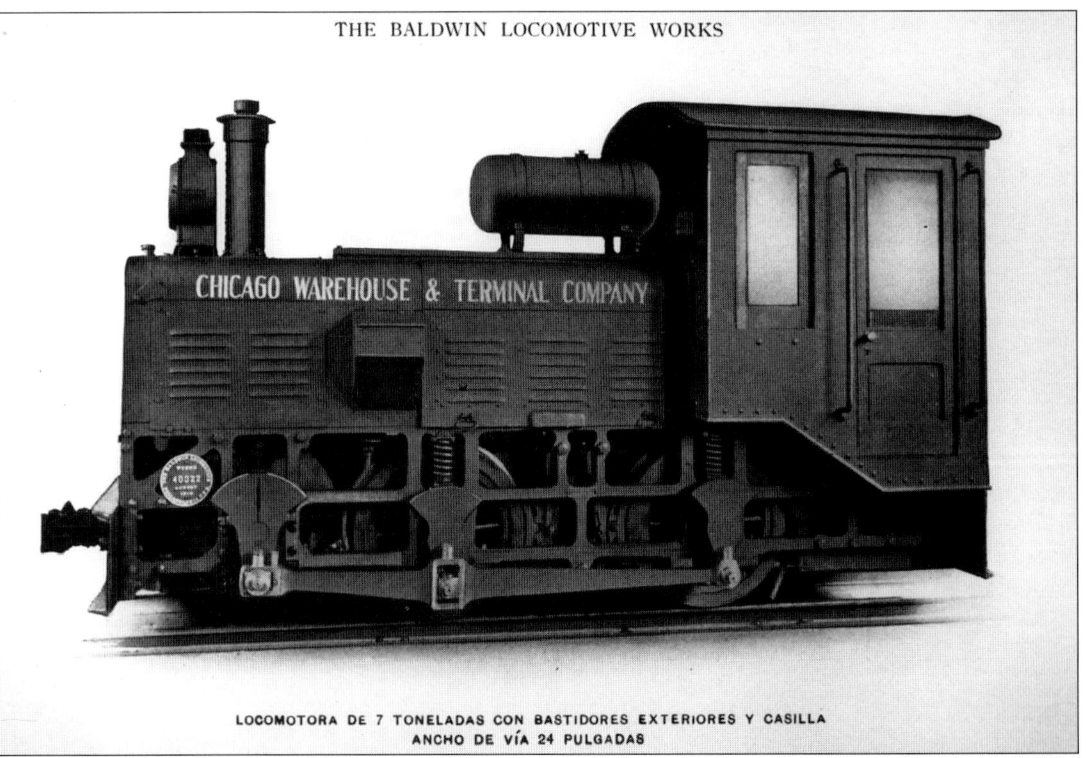
THE BALDWIN LOCOMOTIVE WORKS

LOCOMOTORA DE 7 TONELADAS CON BASTIDORES EXTERIORES Y CASILLA
ANCHO DE VÍA 24 PULGADAS

A Spanish language catalog produced by Philadelphia-based Baldwin Locomotive Works in 1914 included this illustration of one of the internal combustion engines delivered to the Chicago Warehouse & Terminal.

Fred Ash Collection

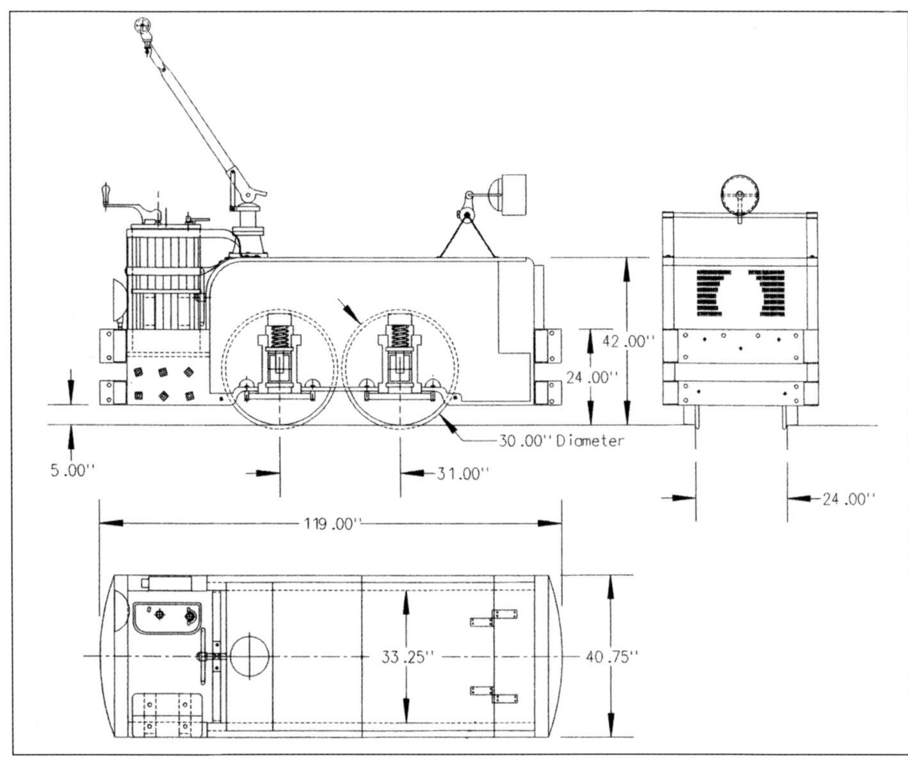

Phil O'Keefe

Scale drawing of Baldwin locomotive 508 prepared by Phil O'Keefe. This locomotive is preserved at the Illinois Railway Museum.

General Electric

During the 1904-1907 period, General Electric supplied 26 2-motor units. Although their original road numbers are not known, they were renumbered 200-225 by 1909. All were scrapped in 1938.

Fleet #	Serial #	Date	Wt.	Model
200-201	2039-2040	1904	10,000	LM103A1
202-205	2080-2083	1905	10,000	LM103C1
206	2074	1905	10,000	LM103D1
207-210	2147-2150	1905	10,000	LM103D1
211-216	2180-2185	1905	10,000	LM103D1
217-220	2194-2197	1905	10,000	LM103D1
221-225	2718-2722	1907	10,000	LM401A1

Western Electrician

A builders photo of one of the 1905-vintage GE's.

Jeffrey

The Jeffrey Manufacturing Company of Columbus, Ohio was a pioneer in the development of electric mine locomotives and accounted for the largest group of units supplied to the system. A 1909 report by Edwin A. Potter, receiver for the Illinois Tunnel Company, listed 68 units numbered 400-467; however construction records list 70. Conversely, a 1958 receiver's report indicates the purchase of at least eight additional locomotives being purchased in 1925, but no corroborating record of this purchase exists. The Tunnel Company classed the Jeffreys as models MH-12 and MH-63, designations that actually referred to the motors (type and class) used.

Originally, the Jeffreys were numbered in the100 and 200 series (70 units) as shown below. Except for a handful of units, surviving roster information is incomplete and somewhat contradictory as to exact fleet number/serial number pairings. By 1909, however, these locomotives had been renumbered into the 400-469 series (68 units). Unfortunately the exact unit-for-unit renumberings are unknown, making it impossible to determine which two units were no longer on the property at the time of the renumbering. Management was evidently pleased with the builder's product, judging by the number purchased during the early years.

Fleet #	Serial #	Date	Motors	Wt.	Jeffrey Class
119	804	08/1904	2-MH12	10,000	26
221-225	868-872	12/1904	2-MH12	10,000	26
?	910-915	03/1905	2-MH12	10,000	26
?	929-936	04/1905-05/1905	2-MH12	10,000	26
?	937	07/1905	1-MH12	6,000	15
?	1013-1024	09/1905-10/1905	2-MH12	10,000	26
162-173	1363-1374	12/1906-02/1907	2-MH12	10,000	25
?	1610-1634	1907	2-MH63	12,000	30

Jeffrey Division, Dresser Industries

The Jeffrey Company photographer was on hand to record this unusual view of six locomotives loaded on a flat car ready to depart the builder's Columbus, Ohio plant.

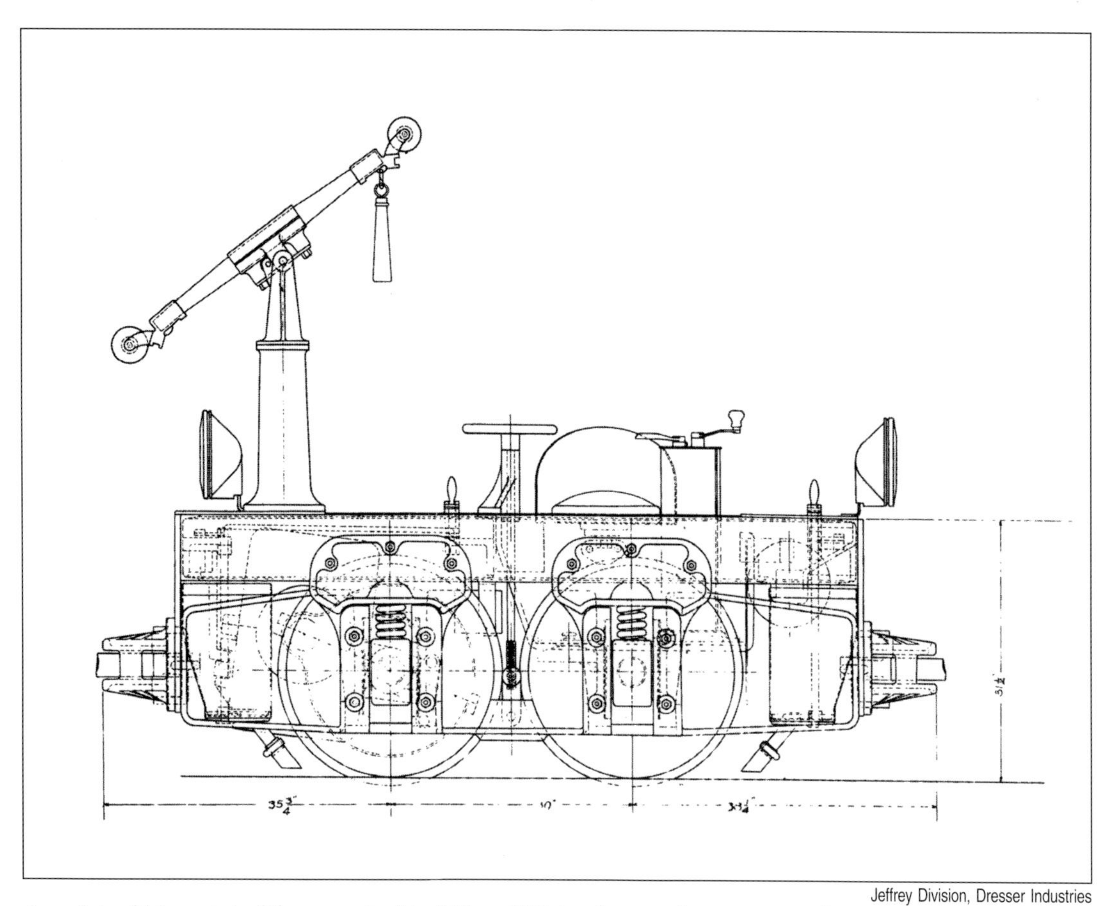

Jeffrey Division, Dresser Industries

An original blueprint of the one-of-a kind Class 15 3-ton locomotive constructed by Jeffery in 1905 and given serial #937. No photographs of this smallest of tunnel electrics are known to exist. Note the "teetering"-type trolley pole.

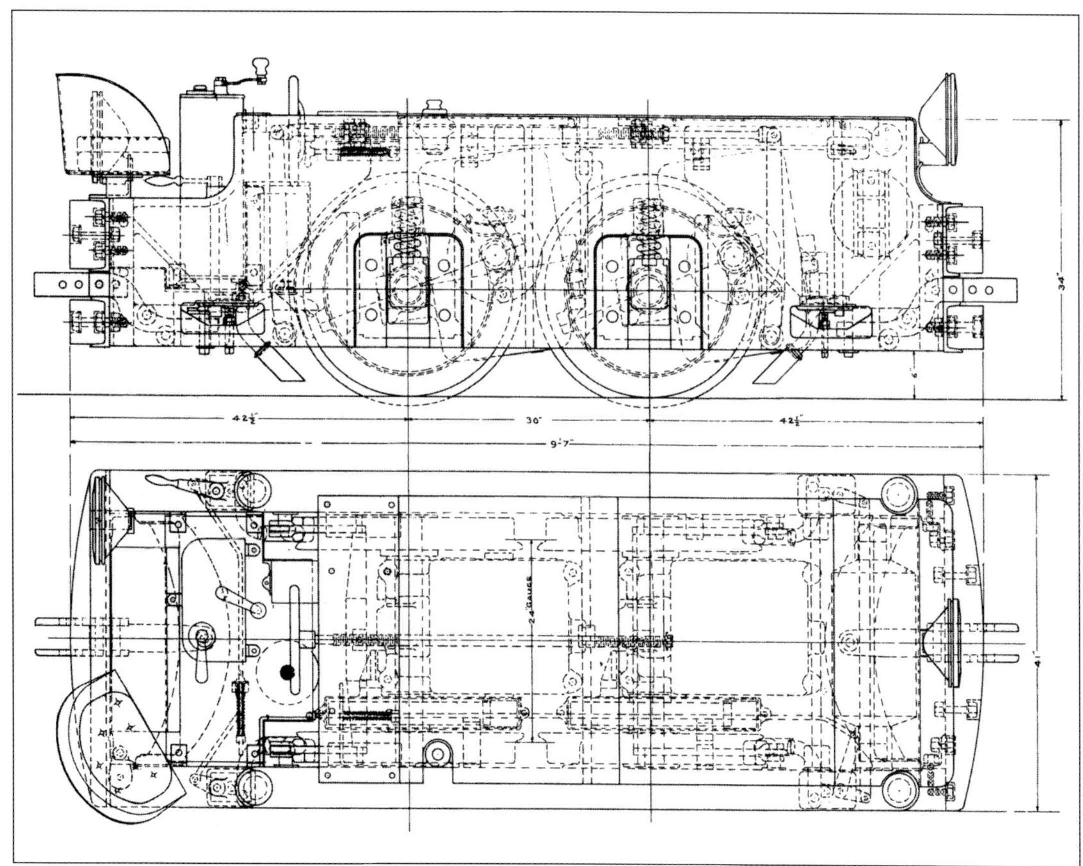

Jeffrey Division, Dresser Industries

Original builder's plan for the Class 25 locomotives; only 12 were built.

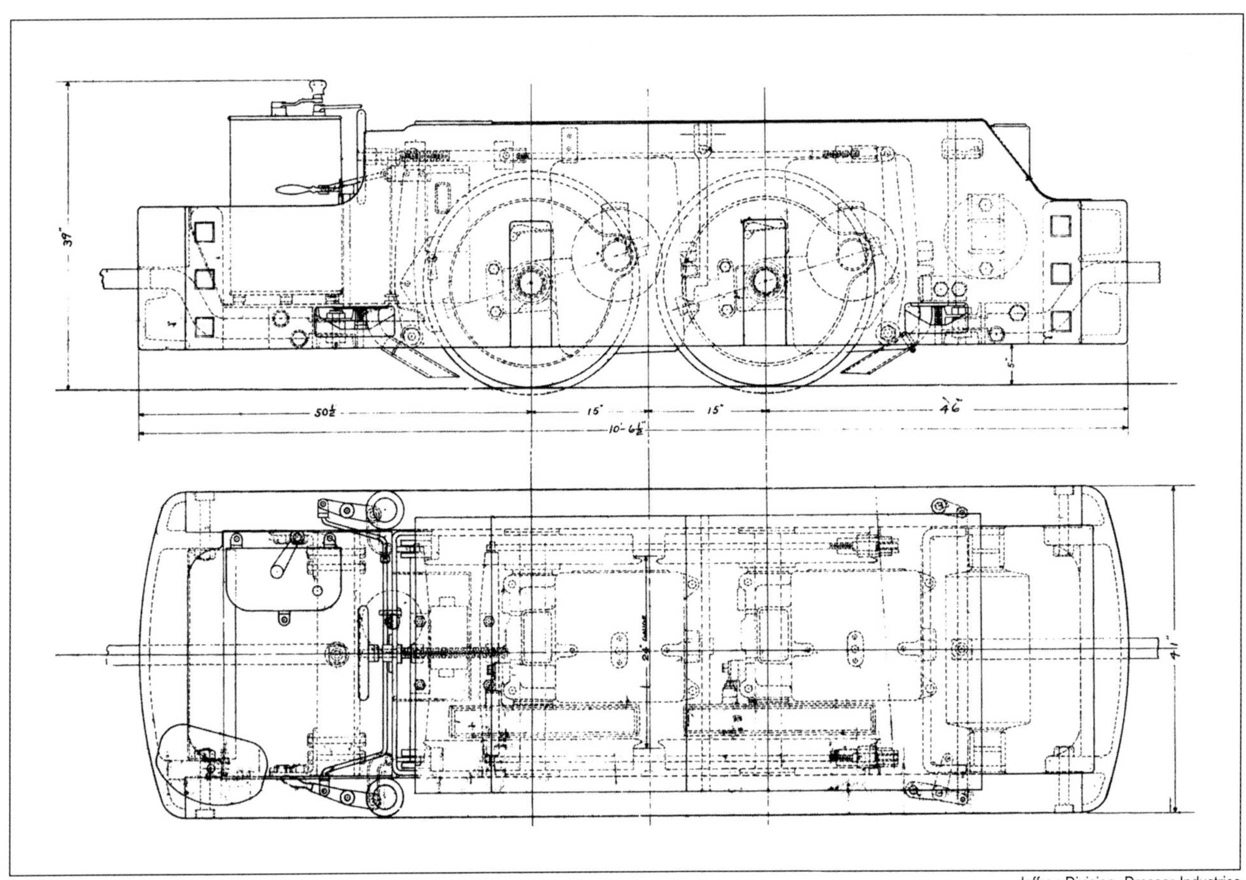

Jeffrey Division, Dresser Industries

Thirty-two of these Class 26 single gondola-style locomotives were produced by Jeffrey to meet the system's motive power needs,

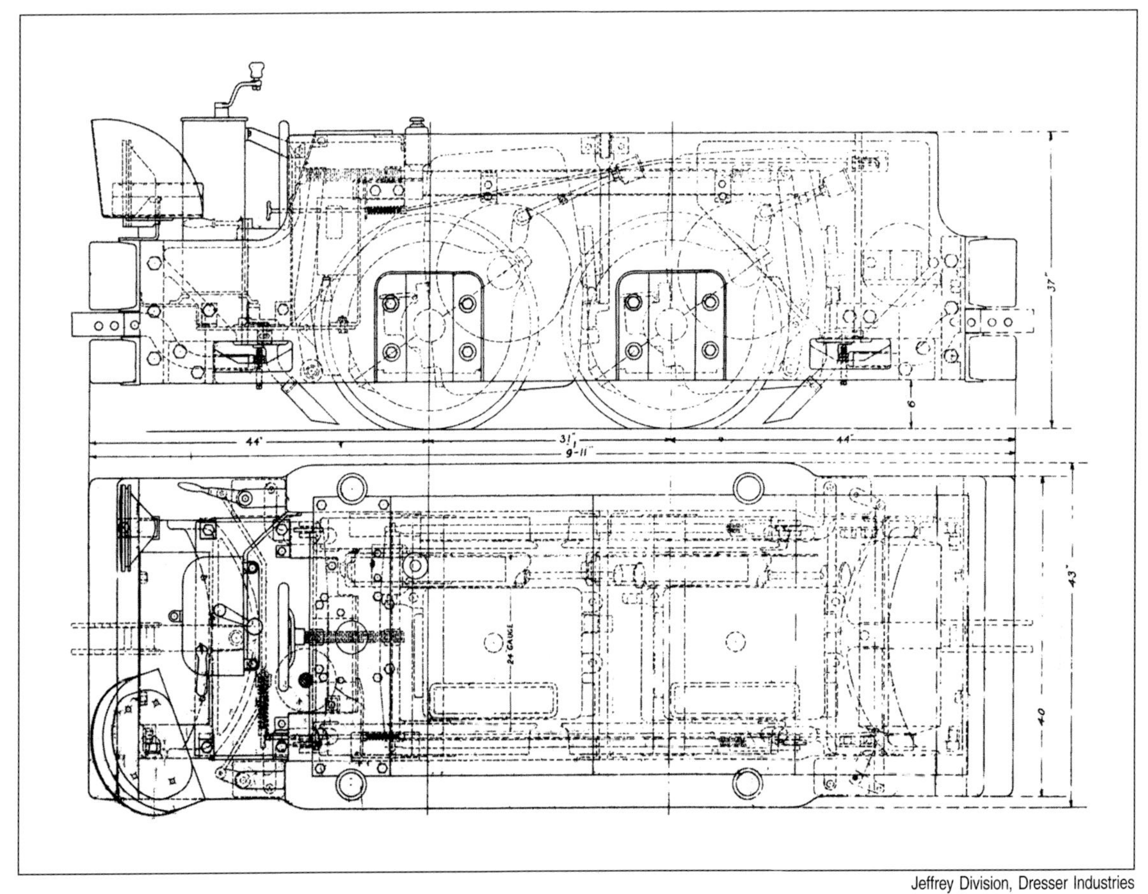

Jeffrey Division, Dresser Industries

The "Big Boys" of the fleet were the class 30 Jeffreys; they numbered 25 and weighed in at 6 tons each.

Jeffrey Division, Dresser Industries

A cut from a 1905-vintage catalog showing a newly arrived Jeffrey locomotive maneuvering an equally new ash car.

Jeffrey-built locomotive #221 was part of a five-unit order built in December 1904.

Jeffrey Division, Dresser Industries

Jeffrey Division, Dresser Industries

Locomotive #162 was among the last 5-ton units that Jeffrey delivered to the Tunnel Company in late 1906 and early 1907.

Jeffrey Division, Dresser Industries

The painters had not yet applied the fleet number to this Class 26 Jeffrey locomotive (serial #1013) when the company photographer photographed it in 1905. The unusual "teetering" trolley pole was apparently a short-lived appliance.

Jeffrey Division, Dresser Industries

Detail of builder's plate on a Class 26 Jeffrey locomotive. Unlike most builders' plates it included the name of the purchasing company (Illinois Tunnel) and its principal official George W. Jackson.

Whitcomb

Based in nearby Rochelle, Illinois, Whitcomb was best known for its line of internal combustion locomotives used by industrial concerns as well as some standard gauge railroads. From 1922 until 1930, Whitcomb also built 46 electric locomotives. Electric locomotive production was then transferred to Baldwin who had acquired control of the company in the interim. Apparently interested in upgrading its motive power, Chicago Tunnel Company purchased the second electric that Whitcomb built. The locomotive's service history and disposition is unknown.

Fleet #	Serial #	Date	Motors	Wt.
?	20002	1922	?	?

Freight & Passenger Equipment

Most of the cars assigned to merchandise, mail and coal service were built to a fairly standardized set of specifications. Generally, a complete car measured approximately 10'6" long, 4' wide and 6'7" high from top of rail to roof (5'1" without the roof section). Average car weight was 3,300 pounds. The minimum turning radius was 15' (20' was the system standard for curves). Although photographs of the first prototype cars taken in 1904 show hand brakes, actual production units came without any brakes at all. These cars were built by the Iowa-based Bettendorf Axle Company and the Kilbourne & Jacobs Manufacturing Company. In the late 1920s and early 1930s, the Chicago Tunnel Company assembled additional cars in its own shops, probably using a mixture of purchased and salvaged hardware.

The high humidity environment of the tunnel caused the cars to deteriorate at an alarming rate. This necessitated the wholesale replacement of most of the merchandise cars and many of the ash cars in the 1920s and early 1930s. After that time, and faced with declining traffic, the company resorted to simply stripping cars for what little useable material remained to be used to keep the rest of the fleet in service. The humidity also frustrated attempts to keep the fleet properly painted. After a while it appears that, with the exception of the car numbers, serious painting efforts were abandoned.

Bettendorf's All-Purpose Freight Car

On June 24, 1904, the *Railroad Gazette* published these views demonstrating W. P. Bettendorf's all-purpose freight car. In a few minutes time the roof section and sides could be added or removed as conditions warranted. The hand brakes shown on this prototype car were not included on the production units and the roof sections were seldom used except for mail and some parcel traffic.

Bruce G. Moffat Collection

Larry Best Collection

The Newman-type spoil cars were built by the Kibourne & Jacobs Manufacturing Company of Columbus, Ohio.

Ed Anderson

A head-on view of ash car 734 which was abandoned at the intersection of Adams and Clinton. Note the metal "house" numbers that were used to identify the ash cars.

Excavation/Ash

Cars used to haul excavated clay and dirt spoil came in a variety of sizes, configurations, and even track gauges. During the first few years of construction a temporary 14" gauge construction railway was used. Animal power was used to move four wheel wooden gondolas to the construction shafts where their contents were transferred to wagons. From available photos it appears that these cars were approximately four feet in length and two feet in width. It is unknown how many cars were built or what their disposition was once they were displaced by the larger two-foot gauge cars.

The two-foot gauge excavation cars essentially consisted of double truck flatcar frames on which were mounted a removable box or a tilt-body gondola. Although little is known about these cars, it is likely that the major components, if not entire cars, came from the Bettendorf Axle Company of Bettendorf, Iowa or the Columbus, Ohio-based Kilbourne & Jacobs Manufacturing Company. Both were major producers of narrow gauge rolling stock for mine railways.

These cars had overall dimensions of approximately 10 feet 7 inches in length and 4 feet in width. The removable box-type cars were designed to be lifted off their flat car base using a derrick. An attached le-

Phil O'Keefe

Phil O'Keefe

Phil O'Keefe constructed this detailed large scale model of an ash car spotted in a typical building basement.

ver allowed the bottom of the floor to swing open, allowing the contents to be neatly deposited in the desired location. As construction activity declined, these cars were gradually reassigned to the handling of heating ash.

The tilt-body version used the "Newman patented dump box" which had a capacity of 3½ yards. The Newman-equipped cars were essentially side dump cars of a type commonly used to carry stone or spoil in mining applications. Unfortunately, these cars were not suitable for ash handling and were retired at a fairly early date. It is possible that their frames may have been reused to build general freight cars as that traffic grew.

Interestingly, all rolling stock was lettered for the parent Illinois Tunnel Company even though they were used, and probably even purchased, by the Illinois Telephone Construction Company.

During tunnel construction a number of specialized cars were on the property to handle the movement of sand, stone, cement and the like. One particularly unusual example was car 9, built by the South Baltimore Steel Car & Foundry Company of Baltimore, Maryland. Having a rather over-built, almost armored, appearance, this double truck steel car had hinged sides and was apparently just under eight feet in length.

Bruce G. Moffat Collection

Within a few years semi-enclosed cars had been introduced.

Merchandise Cars

For merchandise service, the Illinois Tunnel Company had commissioned W. P. Bettendorf, founder of the Bettendorf Axle Company, to design a car that could have broad application on the system. The car that was ultimately designed could probably best be described as a true convertible; one that could be reconfigured in a few minutes to a flat car or an open or covered gondola as shown on page 223.

The basic merchandise car consisted of a steel frame covered with ¼" steel plate, resting on two 4-wheel trucks, and equipped with miniature MCB knuckle couplers. To this basic unit could be added pressed steel side panels to form an open gondola. These side panels were hinged halfway to allow the lower portion to be opened, facilitating rapid unloading of small parcels. It is unclear how often the company's work forces were called on to add or remove these panels given their size and the need to have sufficient space for their storage when not in use. Maintenance of the metal panels and their hinges may have also been a problem.

In addition to the convertible cars, Illinois Tunnel ordered a fleet of flat cars equipped with removable stake sides. These cars were built to slightly different dimensions than the others, measuring 12'7" long, 3'10 ½" wide and, with the stakes added, 5'1" high. Reportedly, these cars also weighed approximately 3,300 pounds and, like their convertible counterparts, they were equipped with MCB couplers. A number of these cars were used in general work service and a few were converted into portable pump cars.

By the 1920s, the company rostered a number of merchandise cars that were equipped with wooden box-type bodies. Surviving newsreel footage shows several of these cars lettered for Mandel Bros., a former Loop department store.

Coal Cars

Outwardly, the coal cars closely resembled the merchandise cars having the pressed steel hinged side panels. However, to unload coal the bodies were designed to tilt, allowing the coal to flow by gravity onto a conveyor for final movement to the customer's coal pile in the boiler room. Few photos of these cars exist.

Chicago Historical Society (ICHi-31231)

What type of cargo was to be handled in this unusually short steel plated freight car can only be left to speculation at this late date. Built in 1903 by the South Baltimore Steel Car & Foundry Company, the car has hinged side panels that would be ideal for unloading coal, construction materials, or other cargo.

Chicago Historical Society (ICHi-31239)

Company general manager and chief engineer George W. Jackson pilots Morgan-built locomotive 109 with car 49 in tow on April 15, 1904. Photographed at an unknown location with a simulated freight load, this car design was ultimately rejected.

Edward Anderson

This wooden merchandise car was left abandoned at the western end of the Jackson tunnel near Halsted in 1959, and managed to survive the 1992 Loop Flood intact as evidenced by this photo taken on October 10, 1992. The stenciled number may have served as something other than a fleet number since it does not agree with the company's 1958 roster.

Bruce G. Moffat Collection

A 1920's newsreel included this image of one of the wooden merchandise cars that were lettered for the long-since defunct Mandel Brothers department store.

Mail Cars

For handling first class mail as well as periodicals and parcel post, Bettendorf designed and built several different types of cars. Although little is known about the mail car fleet, many of the cars were similar in general appearance to the coal and enclosed merchandise cars, having hinged pressed steel side panels to facilitate unloading. Many of these cars had inside frame trucks instead of the usual arch bar or MCB-type trucks. Some cars were also delivered with solid or wire mesh roof panels that could be removed for loading mail. At least one car even had end doors. Based on surviving photographs it appears that the roof panels were not generally used.

For registered mail, the company purchased several cars that were designed with special security features. Because registered mail was frequently used to transport small items of great value such as negotiable securities, the Post Office Department required that the cars assigned to this service have locked compartments to prevent pilferage. The design also had to allow for the continuous monitoring of the contents by postal authorities while in transit To satisfy these requirements, the company purchased several cars that had lockable side and roof panels made of steel mesh.

Chicago Historical Society (ICHi-31229)

The mail cars tended to vary somewhat in outward appearance. This 1905 view of a car with smooth sides was taken on the Grant Park surface trackage. Note the wire mesh roof and hinged end and side panels.

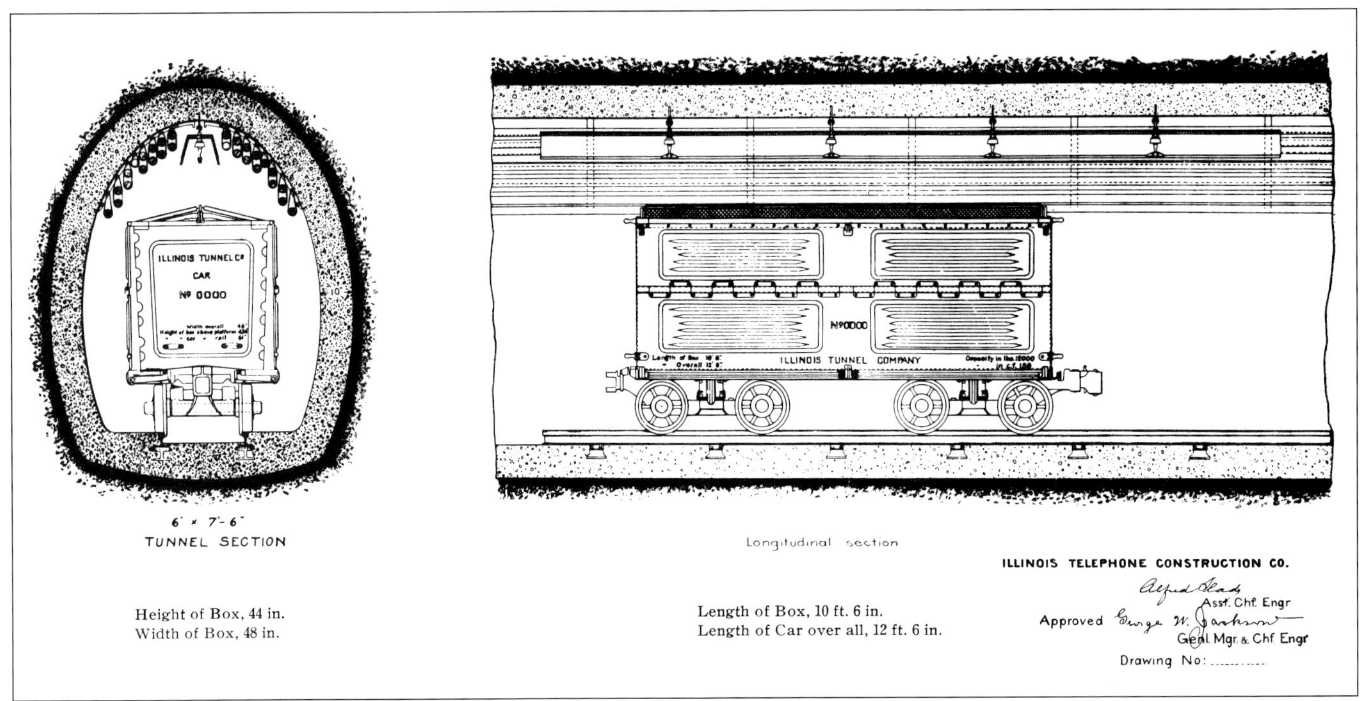

Bruce G. Moffat Collection

Plan of mail car in tunnel. Note the unusual inside frame trucks.

Bruce G. Moffat Collection

One of the registered mail cars at LaSalle Street Station.

Passenger Cars

The company's small fleet of passenger cars was used to take guests on inspection trips through the tunnels. These cars came with either wire mesh or solid metal sides and featured longitudinal (bowling alley) seating. The solid sided cars originally carried the stenciled legend of their builder, the Bettendorf Axle Company. The 1958 receiver's report lists two cars having been built in 1906, however this is probably an error as at least four or five such cars were on the property as early as 1904. At some undetermined date, the company also had a "deluxe" tour car in its fleet. This car had several individual pedestal seats and a chrome-like finish.

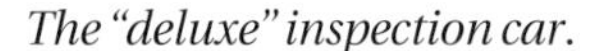

The "deluxe" inspection car.

Bruce G. Moffat Collection

1958 Car Fleet

As part of the receivership proceedings leading to the cessation of operations, the receiver filed this summary of the fleet size as of 1958. At this time only the ash cars and a few of the service cars were still in use. Unfortunately, details concerning renumberngs, builder and year of manufacture were omitted. (It should be noted that a different numbering system was used in the early years as evidenced by photographs showing the mail, excursion and miscellaneous work cars being numbered in the single and double digits.)

Series	Type	# Remaining	Year Built
1-300	Coal	41	1906
500-699	Ash	109	1906
700-799	Ash	56	1929
800-899	Ash	50	1930
900-999	Ash	50	1931
2251-3999	Merchandise	259	1905
5002-5099	Merchandise	93	1925
5100-5199	Merchandise	95	1926
5200-5299	Merchandise	97	1926
5300-5499	Merchandise	187	1927
6000-6499	Merchandise	321	1910
?	Service	18	?
?	Passenger (Excursion)	3	1906
?	Passenger (Excursion)	2	1926

Index

Six months after the last train ran in 1959, workmen installed a bulkhead in the Dearborn Street tunnel adjacent to the Steele-Wedels building. Construction of the new Dearborn Street bridge necessitated the closure.

Eric Bronsky Collection